Time Geography in the Global Context

Time-geography is a mode of thinking that helps us understand change processes in society, the wider context and the ecological consequences of human actions. This book brings together international time-geographic research from a range of disciplines.

Swedish geographer Torsten Hägerstrand is a key foundation for this book, and an introductory biography charts the influences that led to the formation of his theories. A central theme across time-geography research is recognizing time and space as *unity*. Contributions from the Netherlands, the USA, Japan, China, Norway and Sweden showcase the diverse palette of time-geography research. Chapters study societies adjusting to rapid urbanization, or investigate the need for structural changes in childcare organization. The book also delves into green transportation and the interplay between humans and nature in landscape transformation. Applicational chapters look at ICT effects on young people's daily life and methods for engaging clients in treatment practice.

This book situates the outlook for this developing branch of research and the application of time-geography to societal and academic contexts. Its interdisciplinary nature will appeal to postgraduates and researchers who are interested in human geography, urban and regional planning and sociology.

Kajsa Ellegård is Professor in Technology and Social Change, Linköping University, Sweden.

Routledge Studies in Human Geography

This series provides a forum for innovative, vibrant, and critical debate within human geography. Titles will reflect the wealth of research which is taking place in this diverse and ever-expanding field. Contributions will be drawn from the main sub-disciplines and from innovative areas of work which have no particular sub-disciplinary allegiances.

New Geographies of the Globalized World
Edited by Marcin Wojciech Solarz

Creative Placemaking
Research, Theory and Practice
Edited by Cara Courage and Anita McKeown

Living with the Sea
Knowledge, Awareness and Action
Edited by Mike Brown and Kimberley Peters

Time Geography in the Global Context
An Anthology
Edited by Kajsa Ellegård

Space, Grief and Bereavement
Consolationscapes
Edited by Christoph Jedan, Avril Maddrell and Eric Venbrux

The Crisis of Global Youth Unemployment
Edited by Tamar Mayer, Sujata Moorti and Jamie K. McCallum

Thinking Time Geography
Concepts, Methods and Applications
Kajsa Ellegård

For more information about this series, please visit: www.routledge.com/ Routledge-Studies-in-Human-Geography/book-series/SE0514

Time Geography in the Global Context

An Anthology

Edited by Kajsa Ellegård

Routledge
Taylor & Francis Group

LONDON AND NEW YORK

First published 2019
by Routledge

2 Park Square, Milton Park, Abingdon, Oxfordshire OX14 4RN
52 Vanderbilt Avenue, New York, NY 10017

Routledge is an imprint of the Taylor & Francis Group, an informa business

First issued in paperback 2020

British Library Cataloguing-in-Publication Data
A catalogue record for this book is available from the British Library

Library of Congress Cataloging-in-Publication Data
A catalog record has been requested for this book

ISBN: 978-1-138-57378-9 (hbk)
ISBN: 978-0-367-66552-4 (pbk)

Typeset in Times New Roman
by Swales & Willis Ltd, Exeter, Devon, UK

Contents

Figures

Tables

Contributors

Yoshio Arai is Professor of Human Geography at the Department of Human Geography, the University of Tokyo at Komaba. He received his PhD from the University of Tokyo in 1993. His research interests stem from a broad area of human geography with particular emphasis on the time-geography of urban human activities, population geography, the geography of the information society and historical-regional analysis of transportation and telecommunication systems.

Yanwei Chai is Professor of Human Geography at the College of Urban and Environmental Science, Peking University. He obtained his PhD from Hiroshima University in 1994. His research focuses on transitioning Chinese cities, urban spatial structure, space-time behavior and planning, especially from a time-geographic perspective. He introduced time-geography into China's geographical field in early 1990s, and has been leading the Spatial Behavior and Planing research network since 2005 in mainland China and the worldwide Urban China Space-time Behavior research network since 2015.

Martin Dijst, by discipline, is an urban geographer who in 2009 was appointed full Professor of Urban Development and Spatial Mobility at Utrecht University, the Netherlands. On December 1, 2017, he was appointed Director of the Department of Urban Development and Mobility at LISER, Esch-sur-Alzette, Luxemburg. His research, on activity and travel behavior, accessibility, impact of information and communication technologies, exposures to unhealthy environments, social interactions with people, and urban metabolism, is most often positioned in a time-geographical framework. He is also the author of the entry "Time geographic analysis" in the *International Encyclopedia of Human Geography*.

Kajsa Ellegård is Professor in Technology and Social Change, Linköping University, Sweden. She was involved in Hägerstrand's research group in 1974 in a project on public transportation in the future. Her research has since then been inspired by time-geography. She has researched work organization in the automobile industry, the development of the dairy industry in Sweden, the division of labor in households, and how individuals' activity patterns in everyday life influence energy use in households. She has also developed a time-geographic diary method and is the author of the entry "Time-Geography" in *Oxford Bibliographies in Geography* (ed. B. Warf). She organizes an interdisciplinary and international network of resarchers interested in time-geography and arranges yearly meetings in the time-geography community.

Tomas Germundsson is a professor in human geography at Lund University, Sweden. His research has been on landscape changes and the cultural and social history of southernmost Sweden, especially during the 18th, 19th and early 20th centuries. He is also involved in research on present changes in the South-Swedish coastal landscape and the planning process in connection to rising sea levels. He has conducted research on heritage and landscape where he has used time-geography as frame for integrating issues of natural and cultural landscape preservation and also dimensions of past and present.

Eva Magnus is an associate professor/PhD at the Institute of Neuromedicine and Movement Science, at the Norwegian University of Technology and Science in Trondheim, Norway. She is engaged in teaching occupational therapy students. Her research has a focus on disability studies. Her particular interest is in how to develop knowledge of the everyday life of disabled persons, and especially of how restrictions challenge participation in desirable activities and inclusion.

Harvey J. Miller (PhD, Ohio State University, 1991) is the Bob and Mary Reusche Chair in Geographic Information Science, Director of the Center for Urban and Regional Analysis and Professor in the Department of Geography at the Ohio State University. His research focus is at the intersection between geographic information science and transportation science.

Kohei Okamoto is a professor in the Geography Department, Graduate School of Environmental Studies, Nagoya University. He is the author of *Cognition and Behavior in Urban Space* (2000) and editor of several books, including *Traditional Wisdom and Modern Knowledge for the Earth's Future* (2014) and *Integrated Studies of Social and Natural Environmental Transition in Laos* (2014). His research interests are behavioral geography, urban geography and the history of geographical thought.

Carl-Johan Sanglert has a PhD in human geography and has a professional background in the heritage sector. His thesis is based on an analysis of the landscape concept in modern landscape planning and environmental policies. Time-geography plays an important part in the thesis in order to understand the relation between various planning documents and the actual landscape, for instance in the division of the landscape between different kinds of expertise. He is currently working at the county administration in Jönköping county, Sweden.

Ying Song (PhD, Ohio State University, 2015) is an assistant professor in the Department of Geography, Environment and Society at the University of

Minnesota, Twin Cities. Her general interest is GIScience, focusing on applying and developing spatial analytical methods to visualize, explore and investigate movement and change in geographic space with respect to time.

Yiming Tan is an associate research fellow at the School of Geography and Planning in Sun Yat-sen University. Her research focuses on urban social geography and time-geography. She is interested in urban socio-spatial dissimilarity and segregation from the perspective of space-time behavior, especially the impacts of geographic and social contexts.

Eva Thulin, PhD, is an associate professor in human geography at the University of Gothenburg. Her research interests include time-geography, digitalization and how the ubiquitous presence of digital spheres restructures people's everyday lives, their sociality and uses of time and place. Research topics include the leisure-time shifts and relocations occurring in the Swedish population due to digitalization; the recoupling of online sociality and the mounting binds and constraints associated with massive use of mobile media; and the implications of mobile ICTs on the boundaries of work and family life.

Calvin P. Tribby (PhD, Ohio State University, 2016) is a Cancer Prevention Fellow in the Cancer Prevention Fellowship Program, Division of Cancer Prevention, National Cancer Institute, Bethesda, MD. He is a transportation geographer who examines the intersection between transportation systems, public health and the built environment.

Bertil Vilhelmson, PhD, is Professor of Human Geography at the University of Gothenburg, Sweden. His research interests concern human spatial mobility (virtual and corporeal) and its integration with people's activity patterns, their use of time and place, and well-being. Recent research highlights include generational and gendered changes in daily travel, ongoing shifts in teleworking, and how the increased time spent on the internet affects daily life in terms of time, space and sociality. Ongoing projects also concern the role of proximity and slow modes of transport in promoting sustainable accessibility in urban areas.

Yan Zhang is Associate Professor at the Institute of Beijing Studies, Beijing Union University. She obtained her PhD in Human Geography from Peking University in 2012. Her research field focuses on urban social geography and space-time behavior research. She is specialized in understanding the urban transformation process in Chinese cities and its social and spatial implications on the changing everyday lives of individual residents based on time-geography. She has published one monograph and more than 30 journal articles in Chinese and has been the PI of two National Natural Science Foundation of China research projects.

1 Introduction

The roots and diffusion of time-geography

Kajsa Ellegård

> The material world within human reach is altered not by words but by the grasp of the hand. The word-makers are in power, but for their decisions to turn into something more than vibrations in the air, one, some or all people must engage with material things.
>
> (Hägerstrand 2009: 27, my translation)

The citation above captures the core of time-geographical thinking, urging people, whether they are common citizens, decision makers, planners, scientists or in other occupations, to consider the role of the material world when striving for change. Time-geography emanates from the scientific works of the Swedish human geographer Torsten Hägerstrand and his research group. Subsequently, many researchers in the international scientific community have furthered the development. In 2020, fifty years will have passed since Hägerstrand published the pathbreaking article "What about people in regional science?" (Hägerstrand 1970a), which still is frequently cited.[1] In an era where most publications are popular for a short period it is interesting to find out why a fifty-year-old article still gains interest. By presenting examples of current time-geographic research in the global context, this book will show why Hägerstrand's thoughts presented in that article yield such a long-lasting interest. The contributors are researchers in geography and occupational science, for whom time-geography serves as one source of inspiration.

Before presenting the chapters in this book in more depth, a background to time-geography is given. First, there is a short biography of Torsten Hägerstrand which puts his life and scientific work into a societal and geographic context. Thereafter, time-geographical concepts are briefly presented.

Torsten Hägerstrand's work in a societal context

Like many other Western countries, Sweden went through a rapid industrialization and urbanization process during the 20th century, accelerating from the 1930s. Hägerstrand, born in 1916, experienced this process in the flesh. Societal change, then, characterized the country in which he grew up and worked. However, changes in society are not easy to capture for people living in the ongoing processes, even though elements of the changes might be obvious if looked upon one by one. The time-geographic approach is Hägerstrand's effort to provide intellectual and conceptual tools to capture, describe and analyze the evasive phenomena of ongoing change processes in society and nature.

Torsten Hägerstrand grew up in a small municipality in southern Sweden where his father was a schoolmaster. While attending his father's class, as well as at home, he was exposed to the new school subject *home area studies*[2] and the didactics of the Swiss pedagogue Pestalozzi, underlining the importance of starting the learning process with the simple and then going on with building an understanding of increasingly complicated contexts. Home area studies included teaching about phenomena that coexisted in the region where the children lived. The schoolchildren were exposed to lectures about, for example, local plants, animals, buildings, industries, services, landscape and infrastructure. This was intended to enable them to experience and reflect on the combined outcome of all these phenomena in their neighborhood and identify the wider context of which the individual phenomena were parts (Hägerstrand 1983, 2006; Carlestam 1991; Ellegård and Svedin 2012).

Such a contextual approach is foundational for the scientific thinking and works of Torsten Hägerstrand. As a university student, he was disappointed by the specialization in academia, with its disciplines and the lack of communication between them. He found geography to be one of the least specialized disciplines, but with its own problem, the dominant descriptive regional geography school. Hägerstrand was interested in finding the general principles behind what appeared in a region, and was critical of the static descriptions provided by regional geography (Hägerstrand 1983).

As a PhD student, Hägerstrand was sent by his professor to study what happened in a region from which many people had emigrated due to hard times and famine and gone to the USA during the 19th century. His task was to investigate the abandoned houses, supposedly left by emigrants, marked on the map of the study area, Asby parish in Östergötland county. He found from site visits that these houses were located on extremely low-fertility land. Hägerstrand searched for additional sources of information to deepen his investigation and started to study the church registers, wherein the priests had noted important events in the life of each inhabitant in the parish (e.g. birth, family, marriage, moves between dwellings, emigration and death) (Hägerstrand 1947, 1950). From the combination of information from maps, field visits, and the church registers where he could follow the unique individuals over their lifetime,[3] he concluded that the families who had left the abandoned houses were not the emigrants. Instead, the emigrants came from farms located on more fertile land. Even if they were not rich, they had enough to afford tickets. The former inhabitants of the abandoned houses had moved to the houses that were left by people who had emigrated from these slightly better-off farms. From this research, Hägerstrand identified migration chains, which at the time was something new (Hägerstrand 1951, 1962). The research on migration chains combined the two dimensions of space and time in the study of migration, and it was Hägerstrand's first contribution to a more general orientation within geography.

Hägerstrand's next contribution to geography was in the field of innovation diffusion. Here too he combined the time and space dimensions, now following the geographical spread of innovations in Asby parish. This research, resulting in his PhD thesis "Innovation diffusion as a spatial process" (Hägerstrand 1953, English translation 1967), was an influential contribution, not only to innovation research but also to the quantitative revolution in geography, since he used probability and simulation models in the analyses.

His innovative works, both on migration chains and diffusion of innovations, were in themselves a critique of the dominant regional geography school, in which regions were classified from descriptions based on the nature and human activities at the time of the research. Even if Hägerstrand's research, in both cases, was performed with empirical data from one and the same region, he created general knowledge based on the theoretical principles applied on the studied processes in this region.

The above-mentioned research by Hägerstrand gave him a position among the top geographers in the world. The geography department at Lund University, Sweden, where he worked, became attractive among geographers, and many influential international scholars went there for discussions and inspiration.[4]

However, from the mid-1960s Hägerstrand left both the migration and innovation research behind. Instead, he strived to combine and further develop some fundamental and general insights from these studies in order to formulate a theoretically coherent abstract worldview that would be useful for analyzing and explaining processes in the time-space. Concretely, he was interested in how human activities influence the landscape, how resources are exploited and how activities are organized and performed to sustain both human life and nature. Hägerstrand was anxious about the ongoing overexploitation of limited natural resources and life-supporting systems. He was interested in human-induced processes reshaping the landscape and the conditions these give rise to for living species. Such knowledge might help decision makers to take action against nonsustainable development trends (Hägerstrand 1974b, 1976).

The intended approach, then, should be generally applicable and useful for analyzing and understanding the effects of the change processes in which a manifold of different existents and phenomena are involved in their material contexts (Hägerstrand 1988b). The material contexts concern both urban settings and rural areas (landscape in a broad sense) because change processes will influence what happens in both. Hägerstrand hoped that by providing general concepts and a visual language in a coherent approach, researchers might increasingly bridge gaps between disciplines (Hägerstrand 1991).

The endeavor to develop such an approach, which eventually was labeled time-geography, took off when Hägerstrand was awarded a big grant for a research project on the process of urbanization in Sweden.[5] In this period, the urbanization was intense, and there was an urgent need for understanding the process and its impact on people and society. In the research project,

Hägerstrand's research group studied people's daily lives, mainly in urban settings. They observed people's daily outdoor movements and collected activity diaries in order to identify routines and general patterns in persons' daily activities, including transportation.[6] They presented ideas about how to organize activities, like work, services, housing and transportation in urban contexts and utilized time (scheduling) and space (localization) as core dimensions to analyze the fit between the location of such activities and people's daily needs. Normative activity programs consisting of sequences of activities of importance for individuals in their households were suggested. These were tested in urban settings with various locations of workplaces, dwellings and transportation systems (Hägerstrand 1970b; Hägerstrand and Lenntorp 1974). The urbanization process called for urban and regional planning. Partly, the research of Hägerstrand's group was financed and performed within the Swedish governmental programs for urban and regional planning of the 1970s, in which Hägerstrand was much engaged[7] (Hägerstrand 1970b, 1972, 1988a; Hägerstrand and Lenntorp 1974). Additionally, and even more important, the data material was used as the empirical fundament for developing time-geographic concepts and its notation system, which serve as means for theoretical thinking and communicating about people's activities and geographical movements in the time-space.

Time-geographic assumptions and concepts: a brief orientation

The time-geographic approach is based on assumptions that pave the way for its specific concepts. Time-geography also includes a notation system, which is a tool that helps clarify time-space concepts, relations and processes that otherwise, if observed at all, just might seem to be non-related even though entangled (Hägerstrand 1970a, 1974a). The notation system has a visual expression, which provides insights about processes that are very hard to explain with an ordinary vocabulary. Words do not express the time-space extension of processes analyzed with enough precision, and words do not make clear the material dimension of phenomena. The visualizations are useful when investigating contexts wherein many different kinds of phenomena must be coordinated and find room to meet in the time-space (Carlestam 1991). The basic time-geographic assumptions concern ideas about the indivisible individual as a study object, and how to handle time in time-space analyses (Hägerstrand 1970a, 1974a, 1985, 2009), and the most important concepts are: individual, individual path, bundle (two or more individual paths), prism, population, project, constraints and pocket of local order (Hägerstrand 1970a, 1985; Lenntorp 2004).

In time-geography, the concept *individual* is used in a general way, e.g. for things, animals and human beings. Initially, time-geography was criticized for this stance; the critics believed that humans should not be regarded as physical phenomena.[8] Hägerstrand was well aware that humans have specific properties, but he underlined that human beings also have material bodies that take place and that this physical property is not only important, it might even be decisive for what an individual engages in (Hägerstrand 1985, 2009).

The wide denotation of the concept "individual" is based on Hägerstrand's ambitions to create concepts that can be used for general purposes, and in his studies of processes in the time-space he was eager to explain exactly what was denoted by the general concepts, using concrete examples in his texts. An activity, like laying the table, demands coordination of different kinds of individuals in the time-space. The table, tablecloth, plates, glasses, cutlery, and the person performing the activity are examples of individuals involved. This underlines the material presence of both the person and the things used to perform the activity, and by no means implies that the person performing the activities equals the non-living things in other ways. Rather, the person exerts power over the other kinds of individuals and has a plan to lay the table. The non-living individuals do not actively influence the human's activities, but they are important prerequisites, resources, for the person's opportunities to fulfill the activity according to the plan. The ordering of things might affect people's plans and activities. Also, the table can be laid in different ways and the knowledge and skills of the person doing it are decisive for the result. A three-year-old child will lay the table in another way than his mother or the skilled waiter at the restaurant. The influence of persons' different abilities and knowledge indicate that a mere picture of the time-space process is not sufficient for an analysis, but it helps when exploring the process.

From a time-geographic perspective, each individual has the same amount of time every day (they exist for 24 hours) and every individual is, by its physical body, located somewhere. Everyone also has to use all available time each day, albeit by mere existence at some location. The flow of time is assumed to have a constant pace, and individuals' existence can be measured by clock time. This assumption is made for analytical purposes.[9] Most human individuals are aware of time even though they experience it differently. In time-geographically inspired research, humans' subjective ideas about time can be related to the clock time, thereby bringing interesting thoughts and insights to the analysis.

The time dimension has three important parts: past, now and future, as illustrated in Figure 1.1a. The *past* includes what has happened and it cannot be changed. However, what happened at an earlier point in time can be reinterpreted afterwards.[10] The *future* is time to come, and human individuals make plans for the future. However, all opportunities open for an individual in the future are not fulfilled. First, because the future includes many more possibilities than can be realized, and, second, because of risks of collisions with other individuals, with their prisms, including movements and projects. *Now* is the most obvious part along the time dimension, and at the same time the most difficult to grasp. Now is just an instant, and everybody exists in this instant, which in time-geography is regarded as a continuously moving now. In time-geography, accordingly, now is regarded as the constant transformation of future into past and thereby now constitutes the only point in time when actions can be taken and changes made.

In time-geography, an individual is assumed to be an indivisible physical unit at the scale chosen for the study. This is one outcome and generalization from Hägerstrand's migration studies in Asby parish, where he followed unique human individuals over their lifetime. When combining a lot of information about each person, he revealed that the dominant belief among geographers, that the abandoned houses in the study area were left by emigrants, was wrong. He showed that there was no simple correlation between empty houses and emigrants. Instead, by following the unique humans' movements in space over time he identified the general principle of migration chains. The unit of investigation, then, was the physical individual with her lifelong sequential movement chain, rather than, on one hand, the number of empty houses and the number of emigrants on the other. Hence, Hägerstrand's investigation of migration was based on an assumption that the living individual is an indivisible physical unit from birth to death and the non-living individual is indivisible from its construction to its destruction. Hägerstrand later used the concept "continuant" for the indivisible individual in order to underline that there are individuals of different kinds (Hägerstrand 2009).

The indivisible individual and the time dimension based on clock time (with its past, now and future) lay behind the time-geographical concepts and the notation system used to visualize and investigate processes in the time-space. In time-geography, the concept *individual path* is used to follow the sequence of movements in the time-space of any individual, irrespective of whether it is a thing or a person (Hägerstrand 1970a, 1981). The individual path is not just a tool for visualizing movements by individuals involved in processes in the time-space and their relations therein, it is also a way to get to grips with the processual thinking of time-geography. The individual path is visualized by a continuous line in a diagram along two main dimensions: time and space, see Figure 1.1b. The individual path does not show the individual in itself, instead it is the track or protocol of the individual's previous moves in the time-space, revealing the unique movement pattern of that individual. Thereby the individual path is useful for describing and analyzing where an individual has been located in the past and

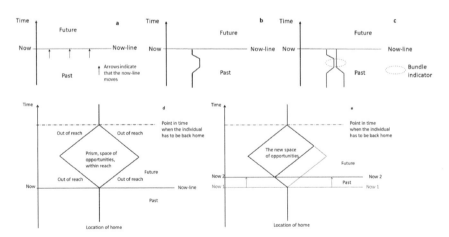

Figure 1.1 Time, with its past, now and future (a); the individual path in the past (b); two individual paths as a bundle (c); the prism principle (d); the changing prism shape as time goes by (e)

until now. When more than one individual is involved, one individual path for each of them is shown (see Figure 1.1c). When the individual paths are located at the same place at the same time they form a *bundle*, which might be used to illustrate collective activities.

But what about the future? As indicated, the future offers more opportunities than a human individual can realize, and the individual has to choose between the opportunities. Time-geography does not predict what choices will be made, instead it reveals what is the space of opportunities or, with another expression, what are the limits of her opportunities. The time-geographic concept for the future space of opportunities for an individual is *prism* (Lenntorp 1976; Miller 1991). The prism is delimited by the individual's geographic location at the now, the maximum speed the individual can move at and the point in time in the future when the individual has to be back at the point of departure, or at another specific place, as illustrated in Figures 1.1d and 1.1e. Then, the prism always opens now, balancing on the now-line. As the now moves upwards along the time-axis, cet-eris paribus, the original prism shrinks, which means that the space of possible location gets smaller. Again, as the now moves upwards along the time-axis, the individual will pass only one of the possible geographical locations at a time, thereby defining the location of the individual path. Consequently, from all the opportunities in a prism, just one can be realized and when now has moved from the bottom of the original prism, the individual path shows how the individual chose to move.

A *population* consists of many individuals of the same kind that exist in a confined region during a delimited period of time (Hägerstrand 1988b, 1972). Then, a population might consist of either e.g. human individuals, birds, flow-ers, tables, which all are material individuals. Within each region, individuals from many kinds of populations that exist together, some of them exert power over others. Individuals in human populations usually control the existence of individuals in other populations. Humans use individuals from other kinds of populations as resources for creating a life as good as possible for themselves. For example, humans use plants and animals as resources, for producing food, clothes and pleasure. Another example is the large-scale use of fossil-fueled cars, which creates freedom for people to move over great distances at an individual level, but has immense environmental effects on, for example, animals, plants, human health and climate change.

The assumption about the indivisible individual brings a novel way to con-sider the relation between the individual and aggregate levels. Hägerstrand's idea was that time-geographic studies should allow researchers to shift from micro to macro level without losing important information about the indivisible individu-als in the shift between the levels. This can be achieved by, for example, using the individual with her continuous activity sequence (e.g. migration history or daily activity sequence) as the unit of analysis also at aggregate levels (Hägerstrand 1972, 1974; Hellgren 2015, Vrotsou 2010).

The time-geographic concept *project* is used for specifying what activities human individuals perform in order to achieve goals that are set up (Hägerstrand

1970a, 1985; Ellegård, 1999). A project, consequently, consists of various kinds of activities that, once they are realized contribute to goal fulfillment. There are individual projects, which relate to goals set by an individual and consist of activities to be performed by the individual herself. There are also organizational projects, which are created by the individuals who are in charge of an organization. In a household with small children, the parents usually set the goals and create household organization projects. At a workplace, the management sets the goals and organizes what activities are to be performed by what employees to achieve the goals of the organizational projects. In both kinds of organizations, the involved persons (children and parents in the household and the management and employees at a workplace) are expected to perform the different kinds of activities that taken together contribute to the goal fulfillment. One and the same activity can be part of more than one project and these kinds of synergies facilitate people's achievement of goals. Also, some projects fail since the individual, or individuals, who should perform the activities meet various kinds of constraints that cannot be overcome.

The time-geographic concept *constraint* is used to specify what hinders human individuals from performing activities in their strivings to achieve the goals of their projects. There are three types of time-geographic constraints: *capacity* (also called capability) constraints, *authority* (also called steering) constraints and, finally, *coupling* constraints (Hägerstrand 1970a; Lenntorp 1976; Mårtensson 1979). Capacity constraints relate directly to the individual's abilities, properties, knowledge and available tools to perform activities. Authority constraints relate to rules, laws, agreements and regulations that are to be followed by individuals in an organization or the society as a whole. Coupling constraints differ from the other two since they overtly include the necessity to couple individuals to each other in the time-space in order for them to successfully perform an activity. For example, in order for a human person to eat, she has to be located at the same place at the same time as the meal is there to be served. In a similar way there are coupling constraints as regards children and adults, e.g. the small child has to be picked up by an adult when the nursery school closes in order for the child to get home and be continuously cared for.

The constraints are interrelated and influence each other. For example, the law says that parents should care for their children, which is an authority constraint giving rise to a coupling constraint. Parents can delegate the care for a while and take the child to a nursery school. Thereby the coupling constraint between the child and the parent is eased while the authority constraint is still there, but a coupling constraint is created between the child and the nursing person. Another example concerns communication technology innovations, which make it possible for people located at different places to communicate, thereby the limits of their capacity to shout or write a letter are overcome. This is a case where the capacity constraints are eased and the technologies help people to overcome coupling constraints.

The time-geographic concept for a place where people regularly perform activities in certain projects, like a home, a workplace, a shop or a city district, is *pocket of local order* (Hägerstrand 1985; Lenntorp 2004; Ellegård and Vilhelmson, 2004).

A pocket of local order is created by people, who set rules for the use of this place and furnish it with things that are important for the fulfillment of the projects to be performed. The order also influences the flow of people (or other individuals) that may enter into the pocket. Homes, factories, offices and shops, then, are ordered in ways that facilitate the specific activities needed for achieving the goals of different kinds of projects, like living a convenient life in the home; producing goods in the factory; supplying services in the shop. The order might change when new conditions appear, like new projects, new people or new ideas. A city district as a pocket of local order is governed by planners and policy makers and it is lived in by its inhabitants who also influence its performance, but it is also used, and influenced, by other people getting there for visits, shopping, work and other activities. The order of a city district as a pocket of local order is planned to facilitate daily life, but the outcome does not always correspond to the plans. Hägerstrand commented about the influence of the material world on the realization of human plans:

> it is very easy to dream up blue-prints for new undertakings but very hard to imagine their fate and their consequences for other legitimate processes when put into practice. Perhaps the trouble is that thought does not encounter in its own world the constraints of space and time.
>
> (Hägerstrand, 1976: 334)

Consequently, the order and material furnishing created can help reveal what kind of pocket of local order it is. The order is upheld by the people involved and can be changed when new needs appear due to new projects of other circumstances. In a family a new order is created e.g. when a baby is born, which also means that at least one new project starts: to raise a child. The new order created is upheld until something happens that calls for a reorganization of it.

These concepts and visualization principles are in various ways and to different extents used by the authors of the chapters in this book, exemplifying the use of the time-geographic approach around the world. It can be seen as an example of geographic diffusion of time-geography as an innovation. Some chapters deal with criticism that over time is directed towards time-geography and suggest how to develop the approach to meet the criticism. Taken together, the authors give creative and constructive contributions to the further use and development of the time-geographic approach.

Use of time-geography in the global context

The contributions to this book are thematically organized. The first two contributions concern the theme of human life in settings influenced by the urbanization process. Thereafter follow two chapters on the communication theme, one concerning physical transportations and the other how information and communication technologies influence daily life activities. The third theme shows the use of time-geography for getting a deeper understanding of meaning, emotions, feelings, experiences in people's everyday life. Finally, there is a theme about the landscape in time-geography, focusing on how that concept is elaborated on in Hägerstrand's writings.

Time-geography: investigating people's living conditions in urban settings

Many time-geographical concepts were developed during the late 1960s in the research about the urbanization process in Sweden (Hägerstrand 1970c, 1990). Then, attention was paid to societal problems emanating from strict partitions between different planning sectors, where, for example, transport systems were planned separately from the location and scheduling of work and services. Time-geography developed conceptual tools to put to the fore the need for coordinating the planning efforts in different sectors. An important proposition was the urge to consider the household as a social unit within which work and household chores are divided between its members regarded as physically indivisible individuals (Hägerstrand and Lenntorp 1974; Ellegård et al. 1977). Largely, this concerns coupling individuals in the time-space so that they can perform activities in projects of importance for themselves, their household and for other organizations they are involved in. Hence, it concerns structural as well as household and individual levels.

Chapters 2 and 3 of the book deal with time-geographically inspired research on the conditions for people in urban settings in two different cultural traditions, Japan and China. Household division of labor, childcare and location of work and housing in cities of the two countries are analyzed, both separately and in comparative studies. Besides presenting empirical time-geographic research, the chapters also give insights about the introduction of time-geography and its development in general in these two countries. In many non-native English-speaking countries, researchers publish at least some works in their native language; this was the case in Sweden with Torsten Hägerstrand and his research group, and with researchers in Japan and China. Therefore, there is limited knowledge in the international English-speaking research community of how the interest in time-geography commenced and grew there. Hence, Chapters 2 and 3 illuminate the diffusion process and use of time-geography in Japan and China.

In Chapter 2 "Time-geography in Japan: Its application to urban life", Kohei Okamoto and Yoshio Arai give the background to the introduction of time-geography to Japan and exemplify how the approach is used. Interestingly, time-geography was introduced to Japan merely as a side effect of a university professor's interest in quantitative methods in geography and his visiting Hägerstrand in Lund to discuss the quantitative approach in innovation diffusion research. Then, in the mid-1970s, time-geography was brought to Japan and young researchers formed a study group to learn more. They found the approach useful for studying the problems emerging from the clash between, on one hand, young dual-income families' need for accessible childcare and, on the other hand, the Japanese work culture with its long working hours. There were long geographic distances between new suburban dwellings and the city center workplaces, causing long commute times. The strong tradition of women being in charge of the children and household chores made it troublesome for women to combine a career, demanding time-consuming commuting, with having children. Wives started to work part time, stopped working after having a child or had no children. Here, the time-geographic coupling constraints provide clues to explain why women's labor force participation is limited. The need for at least one adult to continuously look after children requires parents' presence or nursery arrangements. There is a need for structural changes in the society, and the authors give a time-geographically inspired suggestion of how to solve the problem with parents' accessibility to nursery schools on their way to and from work. The Japanese government recognizes the problems with low birth rates and the shrinking labor force but the measures they suggest are limited to working hour regulations. From a time-geographical perspective, Okamoto and Arai conclude that the government's suggestions "have focused exclusively on the time at work, while they have not considered that time might affect the family and community" (Okamoto and Arai, Chapter 2). Hence, the authors argue for a wider, contextual take to solve the structural problems rooted in the work culture traditions. Solutions based on the micro-level understanding of the daily life of the households gained from time-geographic studies can pave the way for macro-level structural changes.

Experiences of big societal changes in rapidly urbanizing Sweden were one source of inspiration for Torsten Hägerstrand in his development of the time-geographic approach. Big societal changes are currently going on in China, and in Chapter 3 "The time-geographic approach in research on urban China's transition", Yanwei Chai, Yan Zhang and Yiming Tan present how the time-geographic approach was introduced and employed in research on the urbanization in China. Yanwei Chai brought time-geography to China via Japan, where he did his PhD studies in the 1990s, and engaged in the Japanese time-geography research group. Researchers from the two countries made comparative studies of everyday life in cities of different sizes in Japan and China, and concluded that there is a much more even distribution between spouses, both of time spent out of home and working hours, in the Chinese cities than in Japan. The transition of Chinese society from a planned into a more market-orientated economy, however, has far-reaching consequences for daily life in the cities, and the effects differ a lot depending on what type of area people live in. The very organization of the traditional *danwei* compounds, based on close location of workplaces, services, administration, schools and dwellings, made some coupling constraints for people living there relatively weak. However, location of factories in cities close to housing resulted in severe pollution and health problems, and relocation of workplaces is ongoing, which increases the need for longer daily travel to work and affects families' opportunities to spend time together. Modern suburban housing areas located far from workplaces in the city center and industrial districts, force people to commute long distances and increase their use of fossil-fueled cars. This affects the environment, the commuting time and the household activity organization, including the household members' opportunities to jointly perform activities. To some extent, the time-geographic approach is used also for urban planning purposes in China. For example, based on time-geographical principles a tool is developed which produces real-time information about the traffic situation in big cities. It can help people avoid crowded routes. Research on GIS-based geovisualization of people's activities in the urban time-space in Chinese cities reveals the time-space constraints imposed by the built environment. The authors suggest that one development of time-geographic research in China is to include the concepts "project" and "pocket of local order" to find out in more depth what the transition of the society means for the everyday life of people living in various types of neighborhoods in urbanized China.

Time-geography for green transportation and communication with mobile ICT devices

The coupling constraint concept in time-geography deals with the need for individuals to coordinate their activities in the time-space in order to fulfill the goals of the projects they pursue in daily life. Much research concerns how to overcome coupling constraints by improving people's opportunities to come together (couple) at the same geographical location by transportation. Here, the time-geographic concept "prism" is useful and helps reveal differences between people depending e.g. on their access to various kinds of transport means, and their use of the transit network in rush hours. Chapters 4 and 5 have two different takes on how to handle couplings, one by physical transports, and the other by using mobile electronic communication technologies. The chapters discuss two communication-related problems in modern society: environmental effects of transportation are the focus of Chapter 4, while Chapter 5 pays attention to the effects of the increasing embeddedness of ICT in most of the daily activities of young people.

In Chapter 4 " Green, healthy time-geography: Using time-geographic concepts for sustainable mobility planning", Harvey J. Miller, Ying Song and Calvin P. Tribby put to the fore the need for transforming the currently unsustainable mobility system into a more sustainable one, with a focus on favoring positive social, environmental and health effects. The authors suggest a sustainable mobility planning model in contrast to the current conventional planning and refer to results from studies of changes in the transportation system. Two research projects are presented, both aiming at developing methods for transport planning that facilitate sustainable mobility. Methods for estimating expected energy consumption and emissions within prisms are developed, thereby furthering time-geography analytically. Walking is a sustainable and healthy way to move, and in this chapter walkability, built environment and public transit are analyzed by time-geographic principles. These studies show that environmentally important research on daily transit is fruitfully performed by using time-geographic concepts and tools. The more the effects of climate change will appear in people's daily lives, the more important it will be for transit system planners and urban planners to present alternatives to fossil-fueled transport means. Then, one prerequisite is that the daily projects and activity sequences of the indivisible individuals still is performable, and for that purpose the time-geographic approach is inspirational.

Communication with electronic information and communication technologies (ICT) is, at least rhetorically, an alternative to physical transportation. A recurrent question for time-geographers concerns how to handle secondary or simultaneous activities and increasingly so when the use of ICT explodes and permeates most daily activities. In Chapter 5 "Bringing the background to the fore: Time-geography and the study of mobile ICTs in everyday life", Eva Thulin and Bertil Vilhelmson elaborate on this issue taking Swedish young people's mobile ICT use in daily life as a point of departure. They discuss what happens in the daily activity sequences of these young persons when they increasingly use mobile communication technologies as an integrated part of their daily life. The authors contribute to the time-geographic approach by suggesting and testing two concepts for ICT-related mobile communications: foreground and background activities. The young people increasingly and continuously are online, irrespective of what kind of activity they engage in, be it traditional activities (with the mobile ICT device online in the pocket) or ICT-mediated foreground activities. Mobile ICT communication breaks into most activities, demanding immediate re-action in foreground as well as in background activities. The young people develop strategies to manage the technology, which impose two different kinds of problems on their daily life. On one hand, the technology is enabling by offering opportunities to stay involved in processes going on at other geographic locations, while, on the other hand, it causes frictions between mundane activities, like homework, lectures and meals, since the mobile ICT-generated online background activities intermittently call for attention, hence disturbing the ongoing foreground activity. There are also signs of young people getting stressed from situations where they are not in control of such sudden breaks in their activity sequence. The chapter contributes with suggestions about how to integrate suddenly appearing activities of short duration into time-geographic analysis of daily life.

Time-geography: experiences, emotions, health and well-being

In the early days, time-geography, and especially its notation system, was criticized for being physical, not considering human subjectivity, experiences and feelings (Buttimer 1976; Hägerstrand 2006; Giddens 1984; Rose 1993; Baker 1979). Sometimes the notation system was confused with the time-geographic approach as a whole. Time-geographers (Hägerstrand 1983, 1985, 2009; Lenntorp 1976; Mårtensson 1979) argued that time-geography has a materialistic appearance, which is the basis towards which people's experiences and feelings can be related. The notation system, which precisely shows where in the time-space an individual is located, is one component of the time-geographic approach, which should be considered in conjunction with time-geographical concepts, like project, pocket of local order and constraints. The social science criticism does not consider the original intention of Hägerstrand: to create an approach that can handle individuals of different kinds, not only humans, in a similar way based on their mere time-space existence. For Hägerstrand, then, it was important to underline the material dimension of humans and their geographical relations. Hägerstrand's interdisciplinary ambitions with the time-geographic approach were to provide researchers with a way to root studies in the material world, which paves the way for extending the analyses with theories from their own research fields. Chapters 6 and 7 present two different ways to extend time-geographically inspired analysis, with theories from social science in Chapter 6 and occupational science in Chapter 7. The chapters contribute to the discussion about how subjective dimensions can be integrated with and enrich time-geographical analyses.

Chapter 6, "A relational interpretation of time-geography" by Martin Dijst puts to the fore psychological theories of existential feelings and people's relational needs and elaborates on how these might enrich time-geography. Social scientists' criticism of time-geographical representations (the individual path) visualizing an individual's movements in the time-space is the background to Dijst's contribution, where he is inspired by actor-network theory, phenomenology and emotional geography. These theories are used to inform the efforts to extend the time-geographical framework with relational conceptualizations and to get to grips with people's relations and their inner world, thereby complementing the outer world representation of traditional time-geographic notations. The chapter deals with social interaction and emotions in the time-space and suggests new concepts that theoretically pave the way for relational and emotional analyses in time-geographic studies of daily life, like relational string, assemblage of relational strings and embodied exposure. This chapter contributes to furthering micro-level social science use of time-geographic analyses, theoretically informed by psychological research. Insights of this kind may also inform urban and regional planning to create material environments where the existential feelings and relational needs of human individuals are considered.

Chapter 7, "The time-geographic diary method in studies of everyday life", illustrates the use of the time-geographic approach by non-social science researchers. The author Eva Magnus is an occupational scientist, active in the field of medicine and health. Her research is micro level oriented, focusing on the human's abilities (and disabilities) in performing daily life activities in various material, social and geographical contexts. She exemplifies the use of the time-geographic diary method and the constraint concepts in occupational science research and occupational therapy practice in the Scandinavian context. Three types of individual paths are constructed form the diaries, illustrating first a human's activity sequence, second geographical location and movements and third their social companionship in the course of the day. These illustrations deliver background information about phenomena that are hard to capture in traditional treatment situations because of their mundane and evasive nature. Such visualizations based on the client's own diary notes are used as a common ground for discussing with clients what kind of emotions and feelings they experience when they do what activity, where and with whom, and subsequently ideas are developed about what actions can be taken to improve daily life. Thereby, the treatment is grounded on the experiences and descriptions of the client herself, which facilitates the interpretation of what problems and constraints the person meets in her daily life. Magnus also shows that the results from this kind of analysis may serve as arguments for structural changes when occupational scientists engage in discussions with policy makers.

Time-geography and the landscape

Torsten Hägerstrand was not just interested in people, he was interested in how people and other living individuals coexist and use both each other and non-living individuals located in the landscape to sustain themselves. The home area studies from his school years were inspirational for this orientation of time-geography. In his research about the life conditions among the inhabitants in and emigrants from Asby parish, he conducted extensive fieldwork in the landscape. The outmigration of people from the farms led to rapid changes in the landscape. Non-material individuals that previously had played an important role for sustenance on small farms, like tools and buildings, were left behind; they no longer served as resources for humans. Hägerstrand had a deep interest in the landscape and its processual changes due to variations in ongoing activities, and he wrote about it, often in Swedish (e.g. Hägerstrand 1961, 1988a, 1993, 2009). The complex issue concerning understanding of the reshaping of a landscape integrates phenomena like human projects, the constitution of the nature and climate, individuals in the different populations existing therein, societal rules and regulations, previously constructed buildings and other artefacts, which taken together influence the process. It is an ambiguous task to create a scientific approach that can capture all these in a comprehensive way. Hägerstrand's final book, *The Fabric of Existence* (2009) is an effort to argue for time-geography as an ecological approach to analyze processes that influence the ecological and social aspects of sustainability. The ever-changing configurations (bundles) of various kinds of individuals in the landscape, driven by human projects and power, then, are the basis for the processual landscape (Hägerstrand 1993).

In Chapter 8, "What about landscape in time-geography? The role of the landscape concept in Torsten Hägerstrand's thinking", Tomas Germundsson and Carl-Johan Sanglert put to the fore the strivings by Hägerstrand to integrate the landscape into the time-geographic approach. They find that time-geographers have paid limited attention to the landscape concept. Against a biographically inspired background the authors show how Hägerstrand struggled with and developed the landscape concept. The authors underline the importance of the mutual relation between landscape as a view and Hägerstrand's concept of processual landscape. They make clear that the landscape of Hägerstrand is not just a place, it has to do with the scene where individuals exist and relate to each other, from their birth until death, and their activities in a manifold of projects. The landscape, then, is not just a concept for a physical phenomenon, it is where the social and the physical meet and it is always changing due to this. "For Hägerstrand, the landscape thus functioned as a unifying framework in which almost any kind of question or theme could be studied, transcending the established boundaries of academic disciplines and branches of administration" (Germundsson and Sanglert, Chapter 8). The authors underline the dialectic relation between time-geography and landscape in Hägerstrand's thinking and their contribution lays a foundation for further synthesizing landscape research with the time-geographic approach.

Acknowledgements

I would like thank Riksbankens Jubileumsfond for the RJ Sabbatical grant, which made it possible for mer to follow time-geography in the global context. I would like to thank all researchers that took time to discuss time-geography with me during this project: . I would especially like to thank my hosts at the universities visited; Professor Andrew S. Harvey, St. Mary's University, Halifax, Canada; Professor William (Bill) Michelson, University of Toronto, Canada; Associate Professor Dana Anaby, McGill University, Montreal, Canada; Professor Yanwei Chai, Peking University, Beijing, China; Associate Professor Tim Schwanen, Oxford University, Oxford, England; Professor Masago Fujiwara, University of Shimane, Hamada, Japan; Professor Kohei Okamoto, Nagoya University, Nagoya, Japan; Professor Martin Dijst, Utrecht University, the Netherlands, now at Luxembourg Institute of Socio-Economic Research, Luxembourg; Associate Professor Eva Magnus, Norwegian University of Science and Technology (NTNU), Trondheim, Norway; Professor Mei-Po Kwan, University of Illinois, Urbana-Champaign, USA; Professor Harvey J. Miller, Ohio State University, Columbus, USA; and Professor Shih-Lung Shaw, University of Tennessee, Knoxville, USA.

Notes

1 According to Google, by March 14, 2018 this article had been cited 4,014 times, while by April 1, 2012 it had been cited 1,772 times according to Shaw (2012).
2 In Swedish *hembygdskunskap* (Hägerstrand 1983).
3 Or at least until they emigrated or left the parish for other reasons.
4 The department launched the book series "Lund Studies in Geography C, General and Mathematical Geography", which includes contributions by e.g. William Bunge, Edgar Kant, William Garrison, Antoni R. Kuklinski, Richard Chorley and Peter Haggett.
5 In the mid-1960s he had a big grant from the Swedish Riksbank (National Bank of Sweden), which decided to fund research in social science and humanities as part of the bank's celebration of its 300-year anniversary in 1966. This grant laid the foundation for the Research Group in Human Geographic Process and System Analysis at the Department of Geography, Lund University, mostly referred to as the time-geography research group.
6 There are unpublished reports (nos. 3, 17, 38, 39) from this project in the series *Urbaniseringsprocessen* (1969–1970).
7 A substantial part of Hägerstrand's publications stems from works in urban and regional planning and in physical national planning. He also initiated the principle of setting coordinates to all real estate in Sweden in the national register of buildings, which made the administrative geographical borders less of a problem for research and planning.
8 Time-geography developed within a social science, and most social science researchers are primarily occupied with human individuals. When the same concept is used for things and humans, criticisms about physicalism appear (Buttimer 1976, Baker 1979).
9 We do not know if time exists, but we know that it is useful and convenient to measure something we call time by clocks.
10 It can be reinterpreted by people who participated in the event, but also by others. Also, people's memory might result in reinterpretation, since some parts of an event might fall out of memory or appear when something else triggers them.

References

Baker, A. 1979. Historical geography: a new beginning? *Progress in Human Geography*, Vol. 3, 560–570.

Buttimer, A. 1976. Grasping the dynamism of lifeworld. *Annals of the Association of American Geographers*, Vol. 66, 277–292.

Carlestam, G. 1991. Samtal om verklighetens komplexitet, kunskapens och språkets gränser. In *Om tidens vidd och tingens ordning: texter av Torsten Hägerstrand*, ed. G. Carlestam and B. Sollbe. Stockholm: Byggforskningsrådet, pp. 7–18.

Ellegård, K. 1999. A time-geographical approach to the study of everyday life of individuals: a challenge of complexity. *GeoJournal*, Vol. 48, Issue 3, 167–175.

Ellegård, K., and Svedin, U. 2012. Torsten Hägerstrand's time-geography as the cradle of the activity approach in transport geography. *Journal of Transport Geography*, Vol. 23, 17–25.

Ellegård, K., and Vilhelmson, B. 2004. Home as a pocket of local order. *Geografiska Annaler*. Series B: Human Geography, Vol. 86, Issue 4, 281–296.

Ellegård, K., Lenntorp, B., and Hägerstrand, T. 1977. Activity organization and the generation of daily travel: two future alternatives. *Economic Geography*, Vol. 53, Issue 2, 126–152.

Giddens, A. 1984. *The constitution of society: outline of the theory of structuration*. Berkeley: University of California Press.

Hägerstrand, T. 1947. En landsbygdsbefolknings flyttningsrörelser: studier över migrationen på grundval av Asby sockens flyttningslängder 1840–1944. *Svensk Geografisk Årsbok*, Vol. 23, 114–142.

Hägerstrand, T. 1950. Torp och backstugor i 1800-talets Asby [Crofts and cottages in Asby parish during the 19th century]. In *Från Sommabygd till Vätterstrand*, ed. E. Hedkvist. Linköping: Tranås hembygdsgille, pp. 30–38.

Hägerstrand, T. 1951. Migration and the growth of culture regions. In *Studies in rural–urban interaction*, ed. E. Kant. Lund: Royal University of Lund, pp. 33–36.

Hägerstrand, T. 1953. *Innovationsförloppet ur korologisk synpunkt*. Lund: Meddelanden från Lunds universitets Geografiska Institutioner 25. Also published in English as: *Innovation diffusion as a spatial process*. 1967. Lund: C.W.K. Gleerup.

Hägerstrand, T. 1961. Utsikt från Svaneholm. *Svenska Turistföreningens tidskrift*, pp. 33–64.

Hägerstrand, T. 1962. Geographic measurements of migration: Swedish data. In *Entretiens de Monaco en sciences humaines: première session*, ed. J. Sutter. Monaco: Hachette, pp. 61–83.

Hägerstrand, T. 1970a. What about people in regional science? *Regional Science Association Papers*, Vol. 24, 7–21.

Hägerstrand, T. 1970b. Tidsanvändning och omgivningsstruktur. *SOU* 1970:14, bilaga 4. Stockholm: Allmänna Förlaget, pp. 1–146.

Hägerstrand, T, 1970c. *Urbaniseringen: stadsutveckling och regionala olikheter*. Lund: C.W.K Gleerup.

Hägerstrand, T. 1972. Om en konsistent individorienterad samhällsbeskrivning för framtidsstudiebruk, *Ds Ju* 1972:25. Specialarbete till *SOU* 1972:59, Att välja framtid.

Hägerstrand, T. 1974a. Tidsgeografisk beskrivning: syfte och postulat [Time-geographical descriptions: aim and postulates]. *Svensk Geografisk Årsbok*, Vol. 50. Lund: South-Swedish Geographical Society, pp. 86–94.

Hägerstrand, T. 1974b. Ecology under one perspective. In *Ecological problems of the circumpolar area*, ed. E. Bylund, H. Linderholm and O. Rune. Luleå: Norrbottens museum, pp. 271–276.

Hägerstrand, T. 1976. Geography and the study of interaction between nature and society. *Geoforum*, Vol. 7, 329–344.

Hägerstrand, T. 1981. Interdépendances dans l'utilisation du temps. *Temps Libre*, Vol. 3, 53–68.

Hägerstrand, T. 1983. In search for the sources of concepts. In *The practice of geography*, ed. A. Buttimer. London: Longman Higher Education, pp. 238–256.

Hägerstrand, T. 1985. Time-geography: focus on the corporeality of man, society, and environment. In *The science and praxis of complexity*. Tokyo: United Nations University, pp. 193–216.

Hägerstrand, T. 1988a. Krafter som format det svenska kulturlandskapet. *Mark och vatten år 2010*, Stockholm: Bostadsdepartementet, pp. 16–55. In German: Die Kräfte, welche die schwedische Kulturlandschaft formten (1989). *Münchener Geographische Hefte*, Vol. 62, 15–59.

Hägerstrand, T. 1988b. Landet som trädgård. In *Naturresurser och landskapsomvandling*. Report from a seminar on the future. (Based on the lecture: Landskapet som begrepp och förvaltningsobjekt (1988).) Stockholm: Bostadsdepartementet and Forskningsrådsnämnden, pp. 7–20.

Hägerstrand, T. 1990. Urbanization processes. In *Swedish research in a changing society: The Bank of Sweden Tercentenary Foundation 1965–1990*, ed. K. Härnqvist and N.-E. Svensson. Hedemora: Gidlunds bokförlag, pp. 86–104, 453.

Hägerstrand, T. 1991. Tidsgeografi. In *Om tidens vidd och tingens ordning*, ed. G. Carlestam and B. Sollbe. Stockholm: Byggforskningsrådet, pp. 133–142.

Hägerstrand, T. 1993. Samhälle och natur. *NordREFO* 1993:1, 14–59. Nordiska institutet för regionalpolitisk forskning.

Hägerstrand, T. 2006. Foreword by Torsten Hägerstrand. In *By northern lights: on the making of geography in Sweden*, ed. A. Buttimer and T. Mels. Aldershot: Ashgate, pp. xi–xiv.

Hägerstrand, T. 2009. *Tillvaroväven* [The fabric of existence], ed. K. Ellegård and U. Svedin. Stockholm: Formas.

Hägerstrand, T., and Lenntorp, B. 1974. Samhällsorganisation i tidsgeografiskt perspektiv. In *SOU* 1974:2, bilaga 2. Stockholm: Allmänna Förlaget, pp. 221–232.

Hellgren, M. 2015. *Energy use as a consequence of everyday life*. Linköping Studies in Arts and Science No. 662. Linköping University, Sweden.

Lenntorp, B. 1976. Paths in space-time environments: a time-geographic study of movement possibilities of individuals. *Lund Studies in Geography*, Series B, Human Geography, Vol. 44. Lund: Liber Läromedel/Gleerup.

Lenntorp, B. 2004. Path, prism, project, pocket and population: an introduction. *Geografiska Annaler*. Series B: Human Geography, Vol. 86, Issue 4, 223–226.

Mårtensson, S. 1979. On the formation of biographies in space-time environments. *Lund Studies in Geography*, Series B, Human Geography 47. Lund: Lund University, Department of Geography.

Miller, H. J. 1991. Modelling accessibility using space-time prism concepts within geographical information systems. *International Journal of Geographical Information Systems*, Vol. 5, Issue 3, 287–301.

Rose, G. 1993. *Feminism and geography: the limits of geographical knowledge*. Cambridge: Polity Press.

Shaw, S.-L. 2012. Guest editorial introduction: time-geography – its past, present and future. *Journal of Transport Geography*, Vol. 23, 1–4.

Urbaniseringsprocessen. 1969–1970. 56 vols. Mimeo. Lund University.

Vrotsou, K. 2010. *Everyday mining: exploring sequences in event-based data*. Linköping Studies in Science and Technology. Dissertation No. 1331, Linköping University, Sweden.

2 Time-geography in Japan

Its application to urban life

Kohei Okamoto and Yoshio Arai

Introduction of time-geography to Japan

Although previously there had been a few pioneering papers, it was in the 1970s that the impact of the quantitative revolution among English-speaking geographers officially reached Japan. One Japanese advocate of quantitative geography was Teruo Ishimizu, who taught at Saitama University (Figure 2.1). In 1972, he wrote a review paper on quantitative geography and classified its research fields into the following seven categories: (1) point pattern analysis; (2) network analysis; (3) trend-surface analysis; (4) regionalization analysis;

Figure 2.1 Teruo Ishimizu (1932–2005). Professor of Nagoya University, 1976–1995
Reprinted from Department of Geography, Nagoya University (2006)

(5) spatial interaction analysis; (6) spatial diffusion analysis; and (7) spatial behavior analysis (Ishimizu 1972). In the sixth category, Ishimizu introduced Hägerstrand's spatial diffusion studies. Ishimizu and Okuno (1973) provided details of Hägerstrand's (1953) doctoral thesis, "Innovationsförloppet ur korologisk synpunkt", based on the English edition (1967), *Innovation Diffusion as a Spatial Process*, translated by Allan Pred.

In 1974, Ishimizu gained the opportunity to study in Europe by means of a foreign residency research program supported by the Japanese government, and he studied at Lund University for 4 months.[1] After returning to Japan, Ishimizu transferred from Saitama University to Nagoya University, where he published a book entitled *An Outline of Quantitative Geography* in 1976. In this book, Ishimizu added time-geography to the seventh category of quantitative geography indicated above and he explained it as follows:

> Macro analysis [in quantitative geography] is often implemented to examine humans in aggregate rather than individually. Macro analysis is suitable for generalization and theory construction; however, it is difficult to link the analysis results to personal life and problems in an individual's environment. Therefore, it is desirable to begin with micro analysis of individual behavior on the premise that the individual is indivisible and then proceed to an analysis of humans in aggregate (macro analysis).
>
> With respect to this perspective, a new approach called time-geography was initiated in the mid-1960s in the geography department of Lund University, Sweden, based on the work of Professor T. Hägerstrand. This approach is characterized by analyzing the temporal changes in individual spatial behavior, i.e., behavior as a spatial process, while emphasizing the indivisibility of individuals. With the Lund school of time-geography, in addition to Professor Hägerstrand, other members of the department, such as T. Carlstein, M. Jenstav, B. Lenntorp, S. Mårtensson, and E. Wallin, are promoting organized, systematic research.
>
> (Ishimizu 1976: 219–220, translated by the authors)

This is probably the first outline of time-geography to be written in Japanese. In Ishimizu (1976), the author cites the fundamental problems related to consciousness of time-geography; the problems include the indivisibility of the individual, quality of life, microscopic analysis and the time-space mechanics of constraints. Ishimizu also addresses daily path and prism diagrams, which were transcribed from a paper in a journal (Hägerstrand 1968) that could not be obtained in Japan at that time. During his stay in Europe, Ishimizu collected state-of-the-art papers on quantitative geography.

Ishimizu (1976: 223) concluded his explanation of time-geography: "Time-geography has the potential to exert a major influence on quantitative geography in general." In a sense his prediction proved to be correct, since in a quantitative sense, time-geographic studies use big data, such as GPS data, which have become prevalent in this century.

Thanks to the work by Ishimizu, the concept of time-geography became understood among Japanese geographers. Over the 10 years following Ishimizu's study, a few investigations in Japan did adopt such notations as daily paths, however, no full-scale time-geographic research was conducted. An exception was a study by Keiji Kushiya, who was a graduate student of Tokyo Metropolitan University. Kushiya (1985) categorized time-geographic studies into three types: (1) ones that developed the concept of time-geography as a simulation model for formulating regional plans in cities; (2) ones that analyzed the process of structuring society; and (3) ones concerning coexistence between human and natural environments. According to Kushiya, the second and third types are attempts to maintain active aspects of human behavior, since they do not assume that humans are passively dependent on constraints. With respect to the third type of study, Kushiya (1985) conducted an empirical survey of the fishing industry at the mouth of Tokyo Bay. He attempted to define the fishermen's activities as actions that aimed to engage with fish in a time-space prism, which could be transformed by daily weather conditions.

Foundation of the time-geography study group

It was not until the late 1980s that time-geographic studies began to advance in Japan. In 1987, Yoshio Arai formed a time-geography study group, and young geographers (Hiroo Kamiya, Taro Kawaguchi, Satoru Hiromatsu and Kohei Okamoto) participated in it. The young geographers were graduate students of the University of Tokyo and Nagoya University, and they were examining new approaches to research. The study group members began examining published papers dealing with time-geography. They held meetings once or twice a month at the Department of Geography, by then located at the Hongo Campus of the University of Tokyo. Over a period of two years, the group examined almost 50 papers. Based on that examination, the group selected eight papers and translated them into Japanese in order to publish them in Japan.

Table 2.1 lists the eight papers. The first paper is the monumental, influential study by Hägerstrand. The next three papers address family and urban life, which was an area of considerable concern to the study group members. Among them, Pred and Palm (1978) is a chapter in a textbook for undergraduate students, which is both readable and impressive. The fifth and sixth papers deal with modern urban society, which led to today's society. The two last papers address planning and future perspectives.

In January 1989, *Anthology of Time-Geography*, which included those eight papers was published in Japanese. At the end of that year, *Chiri*, a monthly magazine for school teachers and general readers, ran a feature entitled "Time-Geography for the First Time". The feature included an article that applied Lenntorp's PESASP model to the spatio-temporal simulation of nursery schools in a Tokyo metropolitan suburb, and it included also an article that attempted to describe the daily paths of characters in Japanese novels.

The reason for Arai establishing the time-geography study group related to societal conditions in the late 1980s. Under Japan's "bubble" economy of that

Table 2.1 Articles selected for inclusion in *Anthology of Time-Geography* (Arai et al. 1989)

1   Hägerstrand, T. (1970): What about people in regional science? *Papers and Proceedings of Regional Science Association*, 24, 7–21.
2   Pred, A., and Palm, R. (1978): The status of American women: A time-geographic view. In Lanegram, D. A., and Palm, R. eds. *Invitation to Geography*, New York: McGraw-Hill, 99–107 (chap. 7).
3   Forer, P. C., and Kivell, H. (1981): Space-time budgets, public transport, and spatial choice. *Environment and Planning A*, 13, 497–509.
4   Mårtensson, S. (1977): Childhood interaction and temporal organization. *Economic Geography*, 53, 99–125.
5   Miller, R. (1982): Household activity patterns in nineteenth-century suburbs: A time geographic exploration. *Annals of the Association of American Geographers*, 72, 355–371.
6   Pred, A. (1981): Production, family and free time projects: A time-geographic perspective on the individual and societal change in nineteenth-century US cities. *Journal of Historical Geography*, 7, 3–36.
7   Lenntorp, B. (1978): A time-geographic simulation model of individual activity programs. In Carlstein, T., Parkes, D. N., and Thrift, N. J. eds. *Timing Space and Spacing Time*, vol. 2: *Human Activity and Time-Geography*, London: Arnold, 162–180.
8   Ellegård, K., Hägerstrand, T., and Lenntorp, B. (1977): Activity organization and the generation of daily travel. *Economic Geography*, 53, 126–152.

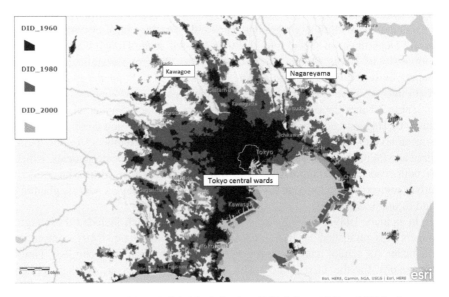

Figure 2.2 Extension of densely inhabited districts (DIDs) from 1960 to 2000 in the Tokyo metropolitan area; DIDs represent urban areas with a population density of 4,000/km² or more, designated by Japan Statistics Bureau. For details, see www.stat.go.jp/english/data/chiri/did/1-1.htm

era, residential housing began moving further away from city centers (Figure 2.2). Long-distance commutes clearly imposed strong constraints on working people's time-space schedules on weekdays. Commuters had to leave home early in the morning and arrived back late in the evening, which, on weekdays, allowed little time to spend with the family. At that time, it was also anticipated that Japanese women would begin to assume a greater social role. Hitherto, they had been traditionally expected to concern themselves largely with domestic matters. Space and time in Japanese people's daily lives were about to change dramatically. Arai considered that a time-geographic approach would be one effective way to identify the related phenomena.

As an economic geographer, Arai also focused on how changes in the use of time could affect consumption activities. In 1987, the Economic Planning Agency of the Japanese government published a pamphlet entitled "Time and Consumption", which provided the following explanation:

> With regard to recent trends in lifestyle patterns among citizens, we find that there is a tendency toward greater elasticity. That conclusion is based on the following phenomena: (1) the number of workers working at times deviating from standard working hours has increased due to the rising importance of the service industries in the economy; (2) because of the increased number of married women in employment, there has been remarkable diversification in the productive hours of married women; and (3) as a result of progressive urbanization, there has been increasing spread of daily activities to the nighttime.
>
> In conjunction with the elasticity in lifestyle patterns, there has developed a need for greater flexibility in shopping hours toward maintaining appropriate levels of consumption. It would be desirable to respond to such needs with flexible hours of commerce.
>
> (Economic Planning Agency 1987: 51, translated by the authors)

Based on their business model, convenience stores were to respond effectively to the expansion of business hours in the commercial and service sectors. Convenience stores originated in the United States in the 1920s. After their introduction into Japan in the 1970s, they rapidly evolved. The number of such stores quickly increased all over Japan in the 1980s, and the number exceeded 40,000 in the 1990s. As of 2017, there were 55,000 convenience stores in Japan. For comparison, we can mention that by then there were 20,000 elementary schools.

Japan's convenience stores are characterized by their long business hours. According to the original model of the 7-Eleven chain in the United States, stores opened at 7:00 and closed at 23:00, which gives 16 business hours. In the 1980s the number of 24-hour stores rapidly increased in Japan. By the end of the 1980s, around 20% of convenient stores operated 24 hours (by 2014, that proportion reached 86%). During the 1980s, the lifestyles of the Japanese changed, and many of their shopping activities advanced towards midnight.

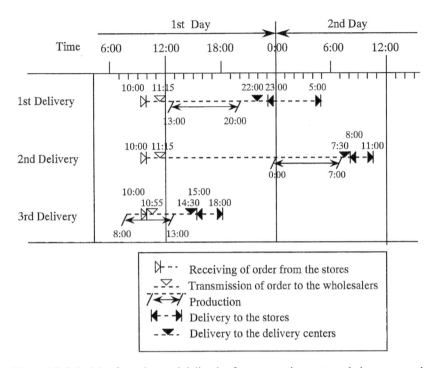

Figure 2.3 Schedules for orders and deliveries for a convenience store chain: an example of fast food in Japan

Reprinted from Arai and Yamada (1994)

Another feature of Japan's convenience stores is the sale of fresh fast food, such as lunch boxes, sandwiches and rice balls (*onigiri*). Convenience stores have limited floor space and store stocks are small. As a result, convenience store chains rely heavily on just-in-time delivery. Fresh fast food is divided into breakfast, lunch and dinner deliveries, and the food is delivered from the food factory to each store via a distribution center at least three times per day. As is evident in Figure 2.3, breakfast is produced at the food factory between 13:00 and 20:00; it is then transported from the factory to the delivery center by 22:00. From the delivery center, the food is transported to each store by 23:00 to 05:00 (first delivery). However, fast food for lunch is produced in the middle of the night (second delivery). Along with the increase in the number of 24-hour convenience stores, there has been a rise in the number of late-night workers – not only in convenience stores but also in food factories and delivery centers.[2]

Empirical research in Japanese cities

Besides publishing *Anthology of Time-Geography*, the time-geography study group undertook empirical research to analyze the daily activities of urban residents in Japan through a time-geographic perspective. Accordingly, the group collected daily activity data of urban residents. Initially, the group undertook a pilot survey in the town of Shimosuwa, Nagano Prefecture in 1988. Subsequently, the group implemented surveys in metropolitan areas to examine the daily lives of suburban residents in Kawagoe, a suburb of Tokyo, and Nisshin, a suburb of Nagoya (Arai et al. 1996, Okamoto 1997).

The surveys in Kawagoe and Nisshin were conducted in 1990. In terms of procedure, a survey request was initially distributed to approximately 10,000 households in each area. Then, among the households that responded positively, around 200 were selected, based on regional distribution, household life stage and employment situation of the wife. The survey sheet was mailed to the selected households, and some face-to-face interviews were conducted when collecting the sheets to address incomplete data compilation.

In those surveys, the space-time budgets for Sunday and Monday for the husbands and wives in each household were examined in detail by means of an activity diary. That diary combined a time-budget survey and travel survey. In the time-budget survey, participants described the details of their activities and individuals who accompanied them in their activities. In the travel survey, participants were asked to provide the details of each trip. Interviews were used to cross-check the two parts of that survey to ensure the accuracy and completeness of the information received.

Figure 2.4 displays the aggregated daily paths of the husbands and wives living in Nisshin, which is located about 15 km from the center of Nagoya. On Monday mornings the husbands left home to go to work, and the figure shows the density of their paths during the morning rush hour. Before 09:00, each path changed to a vertical line, and work began. The figure also reveals that many husbands worked in and around the Nagoya central business district. In the evening, the paths adopt a slanted configuration, indicating the husbands' paths home (Figure 2.4a). By contrast, the wives' paths were more diverse. Many of their paths remained in Nisshin, even in the daytime. Some wives undertook housework, while others had part-time jobs near their homes in Nisshin. A few wives commuted to Nagoya.

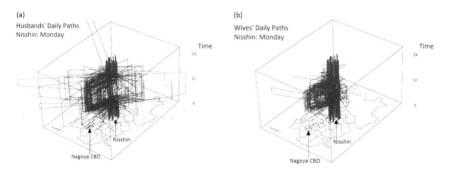

Figure 2.4 Daily paths of (a) husbands and (b) wives living in Nisshin, a Nagoya suburb
Modified after Arai et al. (1996)

Their returning times in the evening were earlier than those of their husbands (Figure 2.4b).

Figure 2.5 shows the spatial distribution of the distance from the home to the workplace for workers living in Kawagoe, Saitama Prefecture, which is about 40 km from Tokyo (Figure 2.2). The horizontal axis in Figure 2.5 shows the distance from home, and the vertical axis plots the cumulative number of workplaces within that distance as a percentage. The cumulative percentage curve for wives amounted to approximately 80% at 7 km from home, which indicates that most wives were working relatively close to their homes. By contrast, the curve for the husbands rose linearly to 40 km, which indicates that the workplaces were evenly distributed over all distance ranges. Thus, the workplaces of the husbands and wives living in those city suburbs differed greatly, revealing gender differences in the spatio-temporal aspects of their daily lives.

In the case of families with small children, the main reason for the differences in the working locations for husbands and wives living in these city suburbs was the difficulty in balancing work and childcare. With respect to that difficulty, it is necessary to understand the culture of long employee working hours and insufficient childcare service that existed in Japan at the time. Table 2.2 shows the average Monday schedules for husbands living in the two areas of Kawagoe and Nisshin. In both locations and regardless of their workplace locations, the average arrival time to be back home for those husbands was after 20:00. At

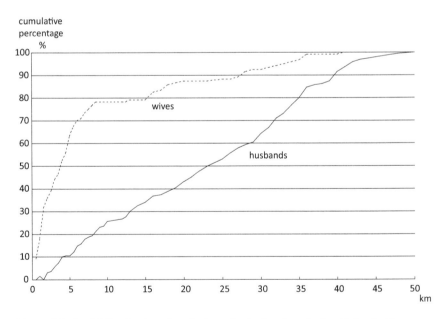

Figure 2.5 Spatial distribution of workplaces for husbands and wives living in Kawagoe, a suburb of Tokyo

Modified after Arai et al. (1996)

Table 2.2 Average Monday schedules for husbands at home and work in 1990 in Kawagoe and Nisshin. Modified after Okamoto (1995)

Survey site	Workplace	Average schedule			N of persons
		leave home/	finish work/	arrive home	
Kawagoe	Tokyo central wards	7:04	18:52	21:15	34
	Tokyo other wards	7:04	18:31	21:05	53
	In / around Kawagoe	7:43	19:08	20:11	74
Nisshin	Nagoya central wards	7:37	19:03	20:44	37
	Nagoya other wards	7:34	18:58	20:21	41
	In / around Nisshin	7:51	19:10	20:04	35

Tokyo central wards: Chiyoda, Chuo, Minato

Nagoya central wards: Naka, Nakamura, Higashi

that time, the closing time of nursery facilities was 18:30 in Kawagoe and 18:00 in Nisshin. Thus, the husbands were unable to pick up their children at those facilities, and therefore the wives had to undertake that role in both places. The working wives had to rush to those the facilities before they closed, and so many wives with young children could take only part-time jobs near their homes.[3]

The difficulty in using nursery facilities for couples living in Japan's metropolitan suburbs can be expressed using the path diagrams of time-geography. In this context, time-geography gives rise to a simple question: if a workplace is distant from home, which is more useful – a nursery close to home or a nursery near the workplace? One answer is a nursery near the workplace since that offers more possibilities. As Figure 2.8a shows, by using a nursery near the workplace, the start and end times of work fit well with nursery service hours. If a child should suddenly suffer from illness, it is possible for the parent to access the nursery quickly from their workplace. From that perspective, having a nursery within the workplace would be ideal. However, for people working in metropolitan centers, it is almost impossible to use nurseries near their workplaces. They cannot use private vehicles owing to traffic congestion and lack of affordable parking spaces, and commuter trains and buses are crowded during rush hours, which is an argument against taking young children by public transit. Thus, they have to use nursery facilities near their homes. However, the service hours of nurseries in the 1990s did not accommodate the long commuting and working hours of city commuters.

Among the survey informants, very few wives worked full time in the metropolitan centers as a result of having small children. For example, one wife living in Kawagoe worked at a central government office in Tokyo. In the morning, her husband drove their 1-year-old child to the nursery and thereafter his wife to the Kawagoe train station on his way to work. In the evening, however, neither the wife nor the husband could reach the nursery before its closing time. So the parents asked a neighbor who used the same nursery for her own child to look after their child at her home until the wife arrived at the neighbor's home after 19:00.

Based on this complex arrangement, it was possible for the family to balance parenting and full-time employment for the wife.

In the Nisshin survey, there was one household in which both the husband and wife worked in the Nagoya city center. Their manner of commuting was rather unusual. In the morning, the husband drove the child to his parents' home in Nagoya, and from there he took the subway to his workplace. The wife took the train and subway to her workplace in the morning, and after work she stopped at her husband's parents' home and took their child back in the car. That way of commuting was possible because the husband's parents' home was located between Nisshin and the Nagoya central business district, and there was a parking space available. However, households with such fortunate transit connections are extremely rare.

In Japan, when the lifetime employment system was formerly the norm, it used to be difficult to return to full-time work after leaving employment. At the same time, social acceptance of parental leave was low, and still today it is not high. Thus, in 1990, wives who lived in the two studied suburbs and hoped to continue full-time work in the cities had to balance work and childcare in intricate ways, as indicated above. In Figure 2.6, the female employment rate by age-group in 1985 shows a typical M shape, indicating that many women left employment as a result of childbirth and childcare. The female employment rate rose from around the age of 40 years, when the burden of child rearing fell; however, following that, most women were in part-time employment, which was different from their situation before childbirth.

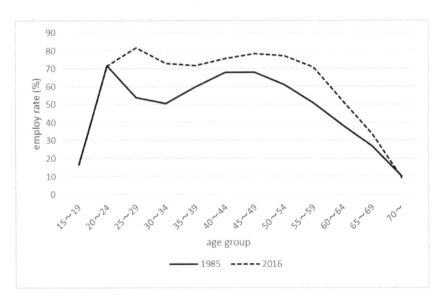

Figure 2.6 Female employment rate by age-group in 1985 and 2016 (whole of Japan)

Source: The labor force surveys (Ministry of Internal Affairs and Communications)

Comparative study on the daily activities in Chinese and Japanese cities

Yanwei Chai was born in China and obtained a doctorate at Hiroshima University in Japan. In his doctoral thesis, he undertook a comparative study between Lanzhou in China and Hiroshima in Japan, from a time-geographic perspective (Chai 1994). Chai transferred to Peking University in 1994, where he advanced his time-geographic research. He investigated the daily activities of urban residents in Dalian in 1995, Tianjin in 1997 and Shenzhen in 1998. The survey method Chai adopted was almost the same as that used in Kawagoe and Nisshin in Japan by Arai et al. (1996). Chai collected activity diary data for married couples for 48 hours from Sunday to Monday. Chai and the present authors compared the data obtained in both countries and jointly examined the differences in the daily lives of urban residents in Japan and China (Arai et al. 2008).

Table 2.3 shows the number of trips and out-of-home activity time per capita on Mondays and Sundays for the husbands and wives living in the three Japanese and three Chinese cities. A trip is defined as moving from one location to another, and an out-of-home activity signified the time spent outside the home. What is especially noteworthy in Table 2.3 is the indicators for Mondays, which show that the spouse difference was remarkably large in Japan, but in China it was much smaller. In Japan, the number of trips made by wives was considerably greater than those by husbands, however, there was little difference in China. Regarding out-of-home activity time, Japanese husbands recorded over twice the times for wives, but there was not much difference in China. On Sundays, the difference between husbands and wives became small in Japan, but it was still greater than in China.

It was evident in the three Japanese urban areas that, on Mondays, the number of wives' trips was greater than those of their husbands. On weekdays, the number of trips made by Japanese husbands was small because they mainly went only to work, while the number of trips by the wives was large because they had to conduct various out-of-home activities related to housekeeping and child rearing – as well as work if they were employed. Thus, the daily activities of urban residents in those three areas of Japan clearly reflected the gendered culture of Japanese society.

On Mondays, the Japanese husbands' out-of-home activity time was 1.3 to 1.5 times longer than that of the Chinese husbands. This reflected the long working hours of Japanese men and the fact that some husbands stopped on the way home to eat and drink before getting home very late. The Japanese wives' out-of-home activity time on Mondays was shorter than that of the Chinese wives. That was because almost half of the wives surveyed in Japan were full-time housewives, and most of the working wives were short-time workers with part-time jobs. The proportion of not employed wives in China was 18% in Shenzhen, 7% in Dalian and 4% in Tianjin.

In China, there were also small gender differences with respect to the workplace. Figure 2.7 presents the distribution of the distance from home to the workplace for husbands and wives in the three Chinese cities in the same way

Table 2.3 Number of trips and total out-of-home activity time per capita in different cities. Modified after Arai et al. (2008)

a) Monday

		Japan			China		
		Shimosuwa	Nisshin	Kawagoe	Dalian	Tianjin	Shenzhen
Husband	N of trips	3.4	2.7	2.7	3.6	3.3	3.4
	Out-of-home activity time (mins)	730.0	758.0	796.9	547.7	541.7	519.4
Wife	N of trips	5.2	5.3	5.3	3.8	3.5	3.4
	Out-of-home activity time (mins)	349.3	375.7	395.7	476.0	462.3	411.3

b) Sunday

		Japan			China		
		Shimosuwa	Nisshin	Kawagoe	Dalian	Tianjin	Shenzhen
Husband	N of trips	3.8	4.6	3.9	3.3	3.3	2.8
	Out-of-home activity time (mins)	423.0	380.6	350.3	251.9	370.0	316.0
Wife	N of trips	4.2	4.6	4.7	3.4	3.3	2.9
	Out-of-home activity time (mins)	272.2	270.5	263.4	230.5	308.3	267.2

Survey years: Shimosuwa 1988, Kawagoe 1990, Nissihn 1990, Dailian 1995, Tianjin 1997, Shenzhen 1998

as Figure 2.5 for Japan. In Kawagoe of Japan in Figure 2.5, the workplace locations displayed a great difference between the husbands and wives. By contrast, in Figure 2.7, the curves for husbands and wives in the same Chinese city were similar, though there were differences between the cities. It is evident that there were virtually no gender differences with respect to workplace locations in China.

To determine why wives in Chinese cities showed a similar employment pattern to their husbands, the present authors and Chai in 2001 and 2002 interviewed operators and users of nursery facilities in Beijing and Shenzhen. We took note of nursing times and meal services. The actual childcare time at these facilities was about 11 hours. One facility served three meals a day to the children.

However, employed married women in Chinese cities are able to rely on support other than from nursery facilities. It was quite probable that the parents of young couples lived together with or nearby and could take care of their grandchildren. In China, the mandatory retirement age was 60 years for men and

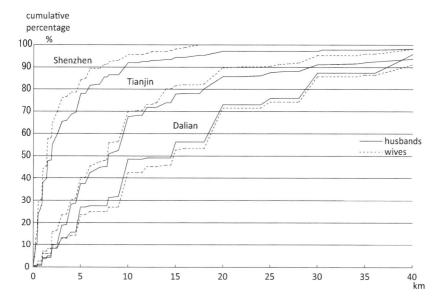

Figure 2.7 Spatial distribution of workplaces for husbands and wives living in three Chinese cities

Modified after Arai et al. (2008)

50 years for women (55 years for women in a managerial position). It is still common for grandparents after retirement to look after their grandchildren. In addition, it was common in China for families to hire people to take care of household chores and childcare in the early 2000s. For example, time-based housekeeping services were available in residential areas of Beijing. Such services were available on a regular or occasional basis, and the rate was 5 yuan per hour on weekdays and 6.5 yuan on weekends at the time of the survey. Some families hired a live-in maid to do housework and childcare. Such housemaids were usually young women from rural areas. One 22-year-old live-in maid from a farming village in Gansu Province worked for a university professor's family in Beijing. She had lived in the professor's house for three years and undertook all the housework, such as cooking, washing, cleaning and childcare. Her wage was 300 yuan a month, and her employer paid for all her meals and other living expenses. In addition, the family covered the cost of her returning home once a year.

From the above-mentioned factors, it is possible to make a comparison of the situation with respect to female employment and childcare, between China and Japan at those times. First, 11 hours of service time for a nursery in China was not long compared with the situation in Japan. The closing time of the nurseries surveyed in Beijing and Shenzhen was 18:00. That was the same as the closing time in Nisshin, and it was earlier than the time in Kawagoe in the 1990s. Therefore, the reason why the use of nurseries in Japan gave temporal constraints to users was not a matter of the service time of nurseries, but rather much longer working and commuting hours in Japan.

Second, regarding childcare assistance other than nurseries, even though it was evident that grandparents who lived together with or lived nearby their children were able to take care of their grandchildren in Japan, this was not as common as it was in China. One remarkable difference between the two countries concerned the gender bias for housework and childcare. In China, it was common for husbands to do housework and childcare, but in Japan in the 1990s, very few husbands did so. Such participation by husbands in Japan is still the lowest among developed countries. For example, the daily time of domestic work for men living with a child aged up to 6 years was 3 hours 21 minutes in Sweden and 1 hour 7 minutes in Japan. The childcare time by men was 1 hour 7 minutes in Sweden and 39 minutes in Japan. As a result, the length of time for housework and childcare among Japanese wives was the longest among developed countries (Eurostat 2004; Ministry of Internal Affairs and Communications, Japan 2011).

Third, regarding housemaids, urban middle-class families often had live-in maids in the early twentieth century in Japan. They were young women from rural villages. At that time, there was a large income disparity between urban and rural areas. However, during Japan's period of high economic growth in the 1950s and 1960s, the increasing demand for labor in the manufacturing industry attracted young women from rural areas to begin working in urban factories. With the decline in income disparity between urban and rural areas, it became difficult for urban middle-class families to hire live-in maids. In Singapore and Hong Kong, many Filipinos work as live-in maids. But in Japan, it is difficult to hire foreigners to do housework because the entry of ordinary workers is severely restricted by immigration control policies.

From the above comparison with China, we concluded that three main factors make it difficult to balance employment and childcare in Japan: long working hours for male workers, deeply rooted norms regarding gender division of labor, and limited possibilities for childcare assistance among working couples.

Development of time-geographic thinking in Japan

As seen above, the time-geography study group focused on the difficulties of balancing employment and childcare in Japan, and it utilized the time-geographic approach to clarify the causes. We attempted to determine what effects time-geography had exerted on geographic academia in Japan. Table 2.4 shows the results of a search we conducted related to changes in the number of articles published in Japan that included the Japanese term for "time-geography". We used two search systems. In search A, we conducted a search for "time-geography" in the titles and abstracts of articles, and in search B, we conducted the search in the whole text. Following the activities of the time-geography study group around 1990, research into time-geography in Japan became most popular around 2000.

Thereafter, the number of studies dealing mainly with time-geography has decreased; however, Japanese academia continues to refer to the subject. In recent years, time-geographic studies have addressed new topics. Of the nine papers published in the 2000s in search A, three applied time-geography to tourism and leisure studies.

One characteristic of Japanese time-geographic studies is the strong connection with nursing problems. In search A, when searching for terms other than "time-geography" contained in the abstracts of articles, the most frequent was "childcare". The search terms and their frequencies were as follows: childcare (9), GIS (geographic information system) (6), city planning (5), female (5), child (5), structuring (3), transportation (3), fishery (3), tourism (3), gender (2), landscape (0), ecology (0), male (0). As the review of Kushiya (1985) indicated, time-geographic studies were originally related to city and regional planning as well as to social theories and ecological studies. In Japan, however, not many time-geographic studies have focused on social theories. The links with ecology and nature have been limited to fishery research.

However, as noted above, early time-geography studies in Japan dealt with the problems of balancing child rearing and employment. Around the turn of the century, there was growing interest in analyzing the accessibility to nursery facilities using GIS and simulation models. Since then, various childcare studies have been conducted, such as into the regional disparity in childcare service provision and local government policies (Miyazawa 2013). In the 2010s, the number of geographic articles dealing with nurseries has increased, though they no longer include the term "time-geography" as a key word.

It can be said that the empirical studies of the time-geography study group around 1990 sparked a rise in childcare studies in Japanese geography. Childcare is now an important topic for geographers interested in Japanese social problems. The Japan Association of Economic Geographers gave its 2017 award to a book entitled *Geography of Childcare and Parenting Support* (Kukimoto 2016).

Table 2.4 Changes in the number of articles written in Japanese that include the term "time-geography"

Period	Search A	Search B
1980–1984	0	1
1985–1989	8	16
1990–1994	6	31
1995–1999	13	55
2000–2004	15	44
2005–2009	7	37
2010–2014	6	28
2015–2017	3	18
Total	58	230

Search A: CiNii articles (https://ci.nii.ac.jp/en)

Search B: J-STAGE (www.jstage.jst.go.jp/browse/-char/en)

In Japan's geographic community, it is widely recognized that economic activities and child rearing are closely related.

Understanding current childcare problems using time-geography

The background for the interest in childcare problems among Japanese geographers is the country's declining birthrate. In 2005, Japan's total fertility rate declined to 1.26. In 2007, the government established the Minister of State for Measures for Declining Birthrate, but still the nation's population has clearly decreased since 2008.

The main cause of the declining birthrate is the progression of later marriages and the decreasing marriage rate. The average age at first marriage in 1990 was 28.4 years for men and 25.9 years for women, which can be compared with 2015 when the figures were 31.1 and 29.4 years respectively. The lifetime unmarried rate was 23% for men and 14% for women in 2015, while in 1990 it was under 5% for both sexes. Furthermore, in recent years, the birthrate of marital couples has declined, and the degree of decline is particularly evident in the Tokyo metropolitan area (Yamauchi et al. 2005). In Figure 2.6, it is evident that the M-shaped trough for female employment rate by age group in 2016 is filling up. That is not because it became possible to balance employment and childcare, instead it was because women were getting married later or remained unmarried, and more women without children continued to work even after getting married.

The difficulty of balancing work and child rearing has continued. In Japan, where childcare assistance is limited, most working couples with young children have to rely on nursery facilities. The number of nursery facilities throughout Japan has grown and the capacity has also increased 1.3-fold from 1990 to 2015. However, many parents cannot afford to send their children to childcare facilities. This has recently become a serious social issue and is known as the "waiting child" problem.[4] The problem is not a matter of the total number of nursery schools and their capacity. Within the same municipalities, there are some filled facilities and some vacant facilities. This problem is a kind of spatio-temporal disparity between nursery facilities and users, and it is necessary to consider this issue from the perspective of time-geography.

Compared with the 1990s, day-care service hours have become longer. In Kawagoe, all nursery schools are open until 19:00, and a few are open until 20:00. In Nisshin, half of the nursery schools are open until between 19:00 and 19:30. Nonetheless, not all nurseries are located at a place that is suitable for parents. If a nursery is located in the opposite direction from the home to the train station used for commuting, there will be difficulties taking children there (Figure 2.8b). Parents desire nurseries located near suburban railway stations. As a result, such schools are popular, and the number of parents who wish to enroll their children there exceeds capacity.

However, it is not easy to construct nurseries in densely inhabited areas. Recently, new social problems have developed regarding childcare centers. People without children or those who have finished raising their children are often opposed to having nursery schools constructed near their homes. They understand the necessity of nurseries, but they do not want them near their own homes. It is a kind of "not in my back yard" (NIMBY) situation. As a consequence, new nursery schools tend to be built in places where not so many people live. Such places are good in terms of natural environment, but they are distant from train stations, and therefore it is difficult for parents to choose a nursery in such a place. That increases the number of "waiting children".

One way to solve this serious problem would be to set up a new kind of nursery service that offers transport services. This would involve establishing a facility called a nursery station, which is located next to a train station. Parents leave their children at the nursery station in the morning and pick them up in the evening. Nursery staff take the children from the nursery stations to remote nursery schools by bus. The children remain at the schools until the evening, when the staff return them to the nursery station, where they wait for their parents to pick them up (Figure 2.8c).

The nursery station is open for a couple of hours in the morning and evening, and so the staff working there are short-time part-timers. With this system, children who cannot be enrolled in nurseries in built-up areas close to train stations can use nurseries with vacancies located outside such areas. This could solve the

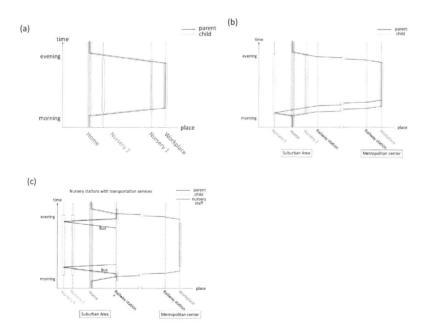

Figure 2.8 Daily paths for parents and children living in a metropolitan suburb

waiting child problem. In addition, compared with regular facilities, parents could take their children earlier in the morning and pick them up later in the evening. Nursery stations with transport services is a time-geographically inspired solution for suburban couples in difficult situation through having to balance work and child rearing.

The nursery station service started in several cities in Saitama Prefecture in the latter half of the 1990s. It is currently being introduced in various local municipalities in Japanese metropolitan areas. Here, we consider the case of Nagareyama, a city in Chiba Prefecture. Nagareyama is located approximately 30 km east of Tokyo city center (Figure 2.2). At one time, urbanization around Nagareyama was delayed owing to poor transport links. However, since the Tsukuba Express railway opened in 2005, the city has been connected to central Tokyo in about 30 minutes. The Nagareyama mayor introduced a slogan, "If you become a mother, you should live in Nagareyama." The mayor did so to attract DEWKS (dually employed with kids) living in Tokyo, and the nursery station system was introduced in 2007. According to an interview survey conducted with city government officials in August 2016, Nagareyama operated the system using two stations and five buses. Each bus went to four or five nursery schools within an hour. About 150 children (4.5% of nursery schoolchildren in the city) used the service in 2015. The city paid 54 million yen (US$500,000) in 2015 to operate the system. Since the system was introduced, the number of people moving to Nagareyama has increased, of which many are in their thirties and with children under the age of 5 years (Table 2.5). This solution has helped parents to ease the balance between employment and child rearing and explains the population influx to Nagareyama. The city has also experienced a rising fertility rate.

Table 2.5 Excess of in-migration over out-migration by age in the city of Nagareyama in 2014

Age	0–4	5–9	10–14	15–19	20–24	25–29	30–34	35–39	40–44	...	Total
N of persons	330	104	24	76	203	507	562	289	152	...	

Conclusion: Japanese society and time-geography

One initial criticism of time-geography was that it was physicalism (e.g. van Paassen 1976). This was due to the path diagrams of time-geography being reminiscent of the movement of liquid particles in the aquarium. Hägerstrand (1989: 3) responded to this criticism as follows, referring to two terms, "meaning" and "matter", in which he uses "matter" to signify everyday materials, such as water, chairs and apples:

Society is not only a set of minds and intangible roles and institutions in interaction . . . society has corporeality. . . . Action in the landscape whatever the meaning is, is also matter acting on matter. Seen in this perspective actions become space-time trajectories of matter. . . . Many will probably say that the matter side of action is self-evident and trivial. I nevertheless believe that our failure to take matter into account has led to our difficulties in judging, for example, the full impact of new technologies and the host of environmental problems, both social, biological and chemical that haunt mankind today.

Thus, Hägerstrand emphasized the importance of the body as matter. However, feminist geographers pointed out the limits of time-geography in terms of considering the body. For example, Rose (1993) made the criticism that, in time-geography, the body shows its position only by its path and that this does not reflect women's corporeality being different from that of men. Rose stated that ultimately time-geography ignores the body. Some feminist geographers assert that time-geography is masculine, and that it prefers to focus on married women's employment issues without considering the ideology that housework is the responsibility of women.

As Harvey (1990) pointed out, the time-geographic path diagram itself does not clearly reflect the social structure that creates constraints upon it. However, what Hägerstrand intended with time-geography was the presentation of an analytic view that connected macro and micro elements. The next stage of the process examines the body as micro and the social system as macro factors.

This chapter has examined the long working hours and traditional gender roles in Japanese society as factors that make it difficult to balance employment and childcare. Japan's society has to change. However, the situation will not change in the short run. There may be many families experiencing difficulties with their lives and in how to manage their time and toward balancing work and child rearing. To raise the quality of life for them, micro measures appropriate for individuals are necessary.

Nursery stations with transport services could be one such measure. In Figure 2.8c, the movement of children from the nursery station to the nursery is represented by a simple line. Some people may raise the criticism that such a time-geographic path does not properly represent the body of the child and makes it look like the movement of baggage. Others may find that the line of the parent does not allow interaction with the nursery itself, which means that parents do not meet nursery school teachers. One of the problems with this service is that there is no daily opportunity for parents to have face-to-face contact with teachers in the nursery where their children spend much of their time. The time-geographic path diagram displays the issues clearly. Rose (1993) made the criticism that the path shows only position, but position is important. To improve people's quality of life, it is necessary to understand how time-geography works and at what scale.

The time-geography study group of Japan recognized that there would be criticisms and controversies related to time-geography, such as those mentioned above. However, in the study group we took the stance of addressing realistic problems from a time-geographic perspective – regardless of such social theoretic discussions. Maintaining that stance consistently has been a major feature of time-geographic studies in Japan.[5] Through that approach, fundamental issues related to Japanese society were eventually identified, which allowed predictions to be made regarding social developments.

In the latter half of the 2010s with the rapid decline in the working population, interest in work-style reform has grown in Japan. The Abe cabinet launched the Council for the Realization of Work Style Reform in 2016, and it issued a provisional action plan. According to that plan, improvement of long-term labor practices is regarded as an important goal to raise the birthrate, increase the number of future workers, and allow more women and elderly people to enter the labor market.[6]

The government campaign calling for reducing long hours at work has led to various political arguments. However, they have focused exclusively on the time at work, while they have not considered how that time might affect the family and community. Regarding the possibility of improving the quality of life for the whole family, it is necessary to consider how the options of all the members' activities can be broadened in the family relationship. In addition, it is important to consider how changes in working hours lead to revitalization of community activities. With these tasks, time-geography will provide valuable analytic perspectives and methods.

Notes

1 Ishimizu handed out a note to people who attended his 60th birthday celebration in 1992. He wrote on the note, "I have wonderful memories of my stay at Royal Lund University in 1974 and spending 4 months with Professor Hägerstrand. Even now, I address his diffusion theory in my lectures at Nagoya University. . . . I clearly remember meeting him again at the International Geographical Congress in Tokyo in 1980 and walking with him near the temple in Asakusa" (Department of Geography, Nagoya University 2006).
2 Nishimura and Okamoto (2001) examined the daily lives of workers undertaking two shifts, day and night, in Toyota automobile factories. For daily lives in a Japanese farming village, see Arai (2002).
3 For details, see Okamoto (1997) and Kamiya (1999).
4 The "waiting child" became a big social and political problem in March 2016. One blog entry, apparently written in mid-February by a woman who had been unable to get her child into a nursery, was titled "Hoikuen ochita Nihon shine!!!" (My child wasn't accepted for nursery school. Japan should die!!!"). When questioned about the entry, Prime Minister Abe stated, "Since it's anonymous, there's no way to tell if it's genuine or not." Angered by this response, a number of people gathered in front of Japan's National Diet Building and held up signs saying, "It was me whose child wasn't accepted for nursery school" (*The Mainichi*, March 7, 2016).
5 When the time-geography study group published its first paper in a journal, the basic time-geographic concepts were summarized as follows: (1) a strong concern with the quality of life and with social welfare, which is reflected in ensuring alternatives for

the handicapped; (2) a tendency to see people not in aggregate but as individuals; (3) a targeted level of analysis intermediate between micro and macro levels; and (4) emphasis upon constraints rather than on demands or preferences (Kamiya et al. 1990). The present authors believe these concepts are still valid and that we should respect them.

6 www.kantei.go.jp/jp/headline/pdf/20170328/07.pdf

References

Arai, Y. 2002. Time and space in the farming village: a time geographical approach. *Japan Review* 14, 195–215.

Arai, Y., and Yamada, H. 1994. Development of convenience store systems in Japan, 1970s–1980s. In Terasaka, A., and Takahashi, S. eds. *Comparative Study on Retail Trade Tradition and Innovation*. IGU Commission on Geography of Commercial Activities, Ryutsu Keizai University, pp.117–126.

Arai, Y., Kawaguchi, T., Okamoto, K., and Kamiya, H. eds. 1989. *Seikatsu no Kuukan, Toshi no Jikan* [Anthology of Time-Geography]. Tokyo: Kokon-Shoin.

Arai, Y., Okamoto, K., Kamiya, H., and Kawaguchi, T. 1996. *Toshi no Kuukan to Jikan* [Space and Time in the City]. Tokyo: Kokon-Shoin.

Arai, Y., Okamoto, K., Tahara, Y., and Chai, Y. 2008. *Chuugoku Toshi no Seikatsu Kuukan* [Activity Space in Urban China]. Kyoto: Nakanishiya.

Chai, Y. 1994. *Comparative Studies on Urban Structure between China and Japan*. Beijing: Peking University Press (in Chinese).

Department of Geography, Nagoya University. 2006. *Geography and Professor Teruo Ishimizu* (in Japanese, no sequential page numbers).

Economic Planning Agency. 1987. *Jikan to Shohi* [Time and Consumption], Tokyo: Printing Bureau of the Ministry of Finance.

Eurostat. 2004. How Europeans spend their time: everyday life of women and men. Data 1998–2002. Available at: http://ec.europa.eu/eurostat/web/products-pocketbooks/-/KS-58-04-998 (accessed June 2018).

Hägerstrand, T. 1967. *Innovation Diffusion as a Spatial Process*. Chicago: University of Chicago Press. Translation of Hägerstrand, T. 1953. *Innovationsförloppet ur korologisk synpunkt*. Lund: C.W.K. Gleerup.

Hägerstrand, T. 1968. On the definition of migration, *Scandinavian Population Studies* 1, 63–72.

Hägerstrand, T. 1989. Reflections on "What about people in regional science?". *Papers of the Regional Science Association* 66, 1–6.

Harvey, D. 1990. *The Condition of Postmodernity*. Oxford: Blackwell.

Ishimizu, T. 1972. Quantitative geography: a theoretical formulation of the geographic space. *Japanese Journal of Human Geography* 24(1), 59–82 (in Japanese with English abstract).

Ishimizu, T. 1976. *Keiryou Chirigaku Gaisetsu* [An Outline of Quantitative Geography]. Tokyo: Kokon-Shoin.

Ishimizu, T., and Okuno,T. eds. 1973. *Keiryou Chirigaku* [Quantitative Geography]. Tokyo: Kyoritsu-Shuppan.

Kamiya, H. 1999. Day care services and activity patterns of women in Japan. *GeoJournal* 48(3), 207–215.

Kamiya, H., Okamoto, K., Kawaguchi, T., and Arai, Y., 1990. A time-geographic analysis of married women's participation in the labour market in Shimosuwa town, Nagano prefecture. *Geographical Review of Japan* 63,766–783 (in Japanese with English abstract).

Kukimoto, K. 2016. *Hoiku, Kosodate-shien ni Chirigaku* [Geography of Childcare and Parenting Support], Tokyo: Akashi-Shoten.

Kushiya, K. 1985. Time-geographic interpretation of fisherman's daily activity on Tokyo Bay, Japan. *Geographical Review of Japan* 58, 645–662 (in Japanese with English abstract).

Ministry of Internal Affairs and Communications. 2011. Survey on time use and leisure activities. Available at: www.stat.go.jp/english/data/shakai/2011/pdf/timeuse-a.pdf (accessed June 2018).

Miyazawa, H. 2013. Geographical studies of welfare issues in Japan since the 1990s. *Geographical Review of Japan* Series B 86, 52–61.

Nishimura, Y., and Okamoto, K. 2001. Yesterday and today: Changes in workers' lives in Toyota City, Japan. In Karan, P. P. ed. *Japan in the Bluegrass*. Lexington: University Press of Kentucky, pp. 96–121.

Okamoto, K. 1995. The daily activities of metropolitan suburbanites and the urban daily rhythm: the case of Kawagoe, a suburb of Tokyo, and Nisshin, a suburb of Nagoya. *Geographical Review of Japan* 68A, 1–36 (in Japanese with English abstract).

Okamoto, K. 1997. Suburbanization of Tokyo and the daily lives of suburban people. In Karan, P. P., and Stapleton, K. eds. *The Japanese City*. Lexington: University Press of Kentucky, pp. 79–105.

Rose, G. 1993. *Feminism and Geography*. Minneapolis: University of Minnesota Press.

Van Paassen, C. 1976. Human geography in terms of existential anthropology. *Tijdschrift voor Economische en Sociale Geografie* 67, 324–341.

Yamauchi, M., Nishioka, H., and Koike, S. 2015. Fertility in metropolitan and non-metropolitan areas of Japan from 1980 to 2000. *Journal of Population Problems* 61(1), 1–17 (in Japanese with English abstract).

3 The time-geographic approach in research on urban China's transition

Yanwei Chai, Yan Zhang and Yiming Tan

Introduction: time-geography and urban transition research in China

Since being introduced into China in the late 1990s, the time-geographic approach and the related space-time behavior research have become influential in China, as urban geographers and planners made sense of the dynamic interactions between individual life experiences on a micro scale and urban social and spatial transformations on a macro scale.

During the last two decades, the time-geographic approach has been applied to studies of Chinese urban space and urban residents' everyday life in China's big cities, and has promoted the rapid development of Chinese behavioral geography, especially the study of space-time behavior (Chai 2013). During the late 1990s to the early 2000s, researchers began to focus empirically on describing the spatial and temporal characteristics of urban residents' everyday activities, and on comparing the urban spatial structure and activity system of several Chinese large cities (Chai et al. 2002). Since 2003, a "behavioral turn" has been recognized in urban geographical research in China, and more and more scholars pay special attention to individual-level spatial behavior, such as residential mobility, commuting, shopping and leisure. Thereby, they tried to reveal the micro process and mechanism of the urban transformation and spatial restructuring of Chinese cities during the transition period (Chai 2005). In 2005 a research network of Chinese urban geographers, urban planners and transportation researchers, labeled the "Spatial Behavior and Planning research group", was formed and it is developing gradually. Based on the time-geographic approach, an increasing number of scholars carry out space-time behavior research and planning practices in large Chinese cities.[1] In addition, time-geography has also been applied to analyze new emerging urban social problems during the market transition, such as the disintegration of work units or *danwei*[2] and its implications on the *danwei* residents' daily life; aging and residential mobility of the aging population; and living conditions of disadvantaged groups emerging during the urban transition, such as migrant workers, the urban poor, and minority groups (Chai et al. 2015).

Since the early 2010s, especially since the Twelfth Five-Year Plan for national economic and social development of the People's Republic of China (2011–2015), China's urban society has undergone a second transformation, focusing on the quality of urbanization rather than the rate of GDP growth and speed of urbanization. Accordingly, the paradigm of urban planning has transformed from "incremental planning", which focused on the development and construction of new districts, to "stock renewal and renovation". There is a change of focus, from the physical and spatial planning to human behavior and social planning, which means a shift from supply-side planning to demand-oriented planning. Furthermore, with the availability of the large volume and high-precision micro-level space-time behavior data collected by, e.g. GPS, cell phone, IC cards, the time-geographic approach has become more and more important and popular, and has been applied to smart city planning practices in some large Chinese cities (Chai and Chen 2018; Zhang et al. 2016; Chai et al. 2014). Therefore, time-geography has a new opportunity for development in its practical application in China.

Aiming at promoting better communication and exchange among urban geographers from different contexts, this chapter provides a critical overview of how the time-geographic approach has been introduced and applied to urban China transition research. We first present a brief history of time-geographic research in China, with a focus on its development stages. This is followed by a review of applications of time-geography to empirical research and planning practices to understand the transformation process and mechanism of Chinese cities. We argue that time-geographic research offers a new perspective for understanding the complexity and diversity of human behavior patterns during the large-scale spatial and institutional transition in Chinese cities. While learning from the West, improvement of data quality and analytic tools has facilitated GIS-based geovisualization of space-time activity patterns in Chinese cities and also improved the understanding of space-time behavior decisions within the constraints of the built environment. Nonetheless, we also recognize that while the techniques and tools used in the visualization of individual path and space-time activity patterns in China have caught up with those in the West, the theoretical development of the time-geographic approach and the empirical studies based on the key concepts of "project" and "pockets of local order" still lags behind.

A brief history of time-geography research in China

In the early 1990s, Chai took the lead in an application of time-geography into empirical studies of residents' daily activities and urban spatial structure of the Chinese city Lanzhou in his doctoral dissertation research[3] (Chai 1996a, 1996b, 1999). In 1996 the National Natural Science Foundation of China sponsored urban geography research, which for the first time adopted a time-geographic framework. Under this project, Chai began to systematically introduce time-geography into China. Now its origin, core concepts and notation system (Chai and Wang 1997; Chai 1998), as well as empirical developments and planning applications of time-geography in Sweden, Europe, the USA, Japan and other countries, were introduced in detail into China (Chai et al. 2000; Chai and Gong 2000, 2001). The potential of applying time-geography to empirical studies of urban China was outlined, for example, on the aging problem (Chai and Liu 2002), and on the geography of enterprises (Liu and Chai 2001).

On this basis, Chai further proposed a research framework of how to understand individual behavior and the temporal rhythm and spatial structure of urban activity systems mainly based on a time-geography approach. He conducted activity diary surveys in three large Chinese cities, Dalian (1995), Tianjin (1997) and Shenzhen (1998), and carried out inter-city comparative studies. These are large cities and they have been experiencing dramatic changes since the reform and opening-up policy from 1978. Furthermore, Dalian and Tianjin reflect the transformation of traditional industrial cities in China, while Shenzhen is a typical representative of the new city formation due to the opening-up policy and rapid urbanization. Most empirical studies were collected in a book titled *The Temporal and Spatial Structure of Chinese Cities* (Chai et al. 2002). From the perspective of everyday life, they concluded that there was no big difference in the spatio-temporal pattern of residents' daily activities among the three cities, and this reflected the fact that the *danwei* system still had profound influence on residents' daily activities during the first decades of market-oriented reform by making up the neat social timetable and by providing job–housing adjacency to their employees. But, in the meantime, we could still find some trends of diversity and complexity of individuals' space-time activity during the transition from the old system to a market economy. For example, the gender difference of time-use and activity pattern in Shenzhen was obviously larger than that in Tianjin (Chai et al. 2002).

As time went by, we continued to follow up and learn from the latest research progress of time-geography. In 2009, there was a review article of recent progress especially in the ten years from 1997 in GIS-based time-geographical research and activity-based research (e.g., Chai and Zhao 2009). In 2010, *Urban Planning International*, a Chinese journal whose readers were both geographers and urban and regional planners, published a special issue themed on "Time-geography and urban planning" co-edited by Sino-US time-geographers Yanwei Chai, Mei-Po Kwan and Shih-Lung Shaw (Chai et al. 2010). There was a translation of Hägerstrand's classic paper of time-geography from 1970 ("What about people in regional science?") in this special issue (Hägerstrand 2010, translated by Zhang and Chai). Besides, in this co-edited special issue, two articles relating to the latest methodological achievements by combining GIS with time-geography were translated, Kwan (2010) and Shaw (2010).[4] The purpose of this special issue was to introduce the time-geographic approach and its latest achievements by combining time-geography with GIS technology into the urban planning research in China. Also, there was a review article of the application of time-geography in planning (Chai et al. 2010) and some thinking and pilot applications of how to apply the time-geographic approach in community planning (Ta and Chai 2010) and tourism planning (Huang 2010).

Since then, a series of articles in edited books on time-geography have been incessantly published in Chinese, for example, a book chapter on Hägerstrand and time-geography was published in the edited book *The Interpretation of Classical*

Thoughts in Geography (Chai and Zhang 2011). The origins of time-geographic thinking back in the early 1940s were introduced, together with the key points from Hägerstrand's classic paper from 1970. The authors' own interpretation of the conceptual framework of the time-geographic approach is presented, including constraint-oriented human behavior analysis, the unity of space and time, as well as the basic principles of individual human beings (Chai and Zhang 2011). The book chapter "Time-geography and its application in urban geography" was published in an edited book, *Thoughts and Methods in Urban Geography*, which introduced several empirical case studies of how the time-geographic approach had been applied in urban context in Japan, the USA and the Netherlands by geographers (Chai and Zhang 2012). For example, this chapter addressed childcare services and their implications for married women's employment in Japan; the simulation of the influence of the urban space adjustment and transportation policy based on the MASTIC model (an improved version of the early PESASP model, developed by Lenntorp in 1976); daily activities and travel behavior of individuals with different family structures in a suburban new town in Zoetermeer in the Netherlands, space-time visualization and geographical calculation of gender differences in the USA; and space-time activity pattern analysis in urban China (Chai and Zhang 2012).

These reviews and introductory articles inspired an increasing number of scholars to get to learn more about time-geography, and researchers carried out diary surveys and applied some concepts or notations of time-geography in their empirical studies on urban space in China from the late 2000s. Yet, mostly due to difficulties of data collection and analysis method, there are only a few research groups and individual scholars, except for Chai and his time-geography group at Peking University, that have so far adopted this approach. Chai and his group continuously launched new rounds of activity diary surveys in Beijing in 2007, 2010, 2012 and 2017, in Shanghai 2017, in Xining 2013, and, from 2010, GPS devices and web-based surveys were added. Zhou and her groups also carried out diary surveys in Guangzhou in 2007 and in 2017, and developed spatio-temporal analysis tools and applied time-geography in empirical studies, e.g. on residential mobility (Gu et al. 2013), and on space-time pattern of daily activity and travel behavior (Zhou and Deng, 2010; Gu et al. 2015) and they analyzed the impacts on urban transportation demand (Zhou et al. 2010) and urban traffic congestions (Gu et al. 2012) in Guangzhou.

Chen et al. (2011), under the supervision of Shih-Lung Shaw and in collaboration with Yanwei Chai, developed a 3D space-time activity analysis tool in a GIS environment based on the activity diary survey data of Beijing in 2007. They carried out the 3D geovisualization of space-time paths and a complex inquiry into space-time paths as well as density maps of subgroups either based on activity or socio-economic attributes. Chen et al. (2011) calculated individual space-time accessibility to urban service facilities under different simulation assumptions (Chen et al. 2015) and put forward a framework of a space-time GIS approach for human behavior studies (Chen et al. 2016). Huang applied the time-geographic approach to tourists' space-time behavior studies, and carried out a diary survey investigating tourists' activities in the Summer Palace in Beijing in 2009, and later on a GPS-aided activity diary survey was conducted for tourists in 2010, again in the Summer Palace in Beijing, and thereafter in the Hong Kong Ocean Park in 2014. With these first-hand data, Huang and her groups studied the space-time behavior pattern of the tourists, the time-budget constraints and subjective satisfaction during tourism, and applied these empirical findings to tourism planning practices (Huang 2009, 2014, 2015; Huang et al. 2016; Huang and Ma 2011; Zhao et al. 2016). Additionally, there were also some small sample case studies using the time-geographic approach to study the living conditions of certain urban social groups in large Chinese cities, for example, for the elderly groups in Beijing (Chai et al. 2010; Zhang et al. 2009; Liu and Chai 2013); for the urban poor in Nanjing (Liu 2005); the migrant workers in Beijing (Lan and Feng, 2010, 2012); for minority groups in Wulumuqi, Xinjiang (Zheng et al. 2009) and Xining, Qinghai (Chai et al. 2017; Tan et al. 2017a, 2017b); and the everyday life of other social groups, such as the middle-class in Guangzhou (Dai et al. 2016) and housewives of Japanese enterprise executives living in Guangzhou (Liu et al. 2010). Besides, Yin (2010) put forward a new development direction, probabilistic time-geography, which combines probability theory with time-geography, and further extended the time-geographic approach by analyzing the uncertainty of individuals' space-time movement. Under the framework of probabilistic time-geography, the classic concepts of time-geography, such as space-time path, space-time prism (potential path area, PPA) and constraints, were redefined, calculated and expressed by the probability concepts, such as random walk and convolution, probability space, sample space (Yin et al. 2015). It laid a new foundation for new quantification and theorization in the time-geographic approach.

Since the late 2000s and early 2010s, the space-time behavior and planning research based on the time-geographic approach in urban China has entered a new stage of increasing and intensifying international communication and cooperation. In addition to collaboration with Japanese time-geographers, we built up further academic exchanges and corporation with time-geographers from e.g. the USA, Sweden and the Netherlands. In April 2011, Chai was invited to give a keynote speech in the special symposium entitled "Space-Time Integration in Geography and GIScience" within the AAG's 2011 Annual Meeting in Seattle, USA. In this symposium, Chai reviewed the development process and the latest achievements of the space-time behavior research and application in urban China (Chai 2013). Additionally, in December 2012, the international workshop on "Space-time behavior and smart travel" was held at Peking University in Beijing, aiming at the application of space-time behavior research to smart travel planning practices which was strongly encouraged by the Chinese government as a key project in the National Science & Technology Pillar Program during the Twelfth Five-Year Plan period. This workshop caused some repercussions at home and abroad, not only because it was in China where the government initiatives of smart city planning made clear the great importance and usefulness of the time-geography and space-time research to guide urban policy and planning practices. It was also very important for Chinese scholars to join the international network of time-geographers. Furthermore, edited conference proceedings were published after the workshop, titled "The frontier of space-time research" in 2014 (Chai et al. 2014). At the AAG's 2015 Annual Meeting in Chicago, USA during the session "Chinese spatial behavior and city planning", Kwan and Chai formally announced the establishment of the international research network of "Urban China Space-Time Behavior" (UCSB).

During this period, more methodological progress was made in data collection and analytical tools for space-time behavior research based on the time-geographic concepts of "path" and "prism". Owing to the increasing scholarly interactions occurring between Chinese and Western geographers, and the rapid development of GIS techniques as well the availability of high-resolution space-time data, the disaggregate-level analysis of space-time paths has achieved major methodological breakthroughs. The space-time activity pattern analysis aided by 3D geovisualization has become very popular in space-time behavior research in urban China (Chai et al. 2015).

More recently, through increasing academic exchanges and communication between China and Sweden, especially with Kajsa Ellegård, the leading scholar of time-geography in Sweden, we are deeply impressed by the latest developments in Swedish time-geography research, which we called the "new" time-geography and we began to reflect on the future development direction of time-geographic research. First, we recognized the unbalanced development of the core concepts of time-geography, especially the concept of "project" which represents the inner side of daily activities, as well as the concept of "pocket of local order" (POLO), which comprehensively represents the context of daily activities and the under-lying power relationship among different actors in a confined place wherein an order is implemented. Therefore, in collaboration with Ellegård, we published a special issue in the Chinese academic journal *Human Geography* and introduced the core concepts of "project" and "pocket of local order" and the most recent applications in the Swedish urban context (Chai et al. 2016). Also, we put forward that the "new" time-geography approach shifted from "out-of-home" to include also "in-home" behavior, from the emphasis on "spatial behavior" with large-scale spatial movements such as migration and commuting, towards the emphasis on "behavior in space and time", which were deeply influenced by social structure and institutional transformation. Second, in terms of theoretical perspective, the "new" time-geography has been experiencing a shift from "absolute space-time" to also recognize "relational space-time". In the view of "absolute space-time", the individual path is treated as a movement from one place to another and along clock time. In contrast, in "relational space-time", the "context" of everyday life experience is emphasized, within which the concepts of "side-by-sideness" clarified the "coexistence" of agents within a concrete place, and where the concept of "before-and-afterness" described the continuous shifting process of "past–now–future". Altogether, the concepts reflected the "relational time" view and were important for our understanding of everyday life (Zhang et al. 2016). Besides, based on the "relational space-time" view, Ellegård and her group created a pro-ject-oriented activity category, a coding system and visualization tools to display the complex relationships between various contexts characterizing individual daily life. This paves the way for new components in the time-geographical notation system for identifying basic events and the formation of pockets of local order. Finally, we argued that "new" time-geography starts to pay special attention to the feedback mechanism and dymanic process of "project–path" dialectics. By underlining the process of "project–activity–project", the time-geographic approach might to a certain extent overcome previous criticisms of neglecting human agency. In future, the time-geographic approach could be applied to and be of great help for understanding the urban transition process and its implications for people's daily life experiences in Chinese cities. We argued that more applications and empirical studies based on time-geography should be made in the fields of the suburbanization of everyday life, the neighborhood-level differentiation of daily activity pattern, domestic labor division and the gender difference of daily activity pattern and so on (Chai et al. 2016; Zhang and Chai 2016; Ellegård et al. 2016a, 2016b, 2016c).

In sum, time-geographic research in China has experienced three developmental stages, and it currently faces a new stage (Table 3.1). In the first stage, during the early 1990s to the mid-2000s, the time-geographic approach was introduced into China, and empirical studies of urban residents' daily activity, and the spatial and temporal characteristics of urban activity system were analyzed. In the second stage, during the mid-2000s to the end of 2000s, especially with inspiration from the USA, more methodological progress was made in data collection and analytical tools for space-time behavior research based on the time-geography concepts "path" and "prism". Space-time behavior research developed rapidly during this period in China, and has become an important approach for urban geography and urban planning practices. Since 2010, motivated by the government and the information technology business, the initiative to apply time-geography-based space-time behavioral research to smart city planning practices has been undertaken. Since 2016, on one hand, we began to realize that the theoretical progress of time-geography research has still lagged behind in China, and the study of space-time constraints and detailed analysis of the contexts within which individuals' activities were encouraged. On the other hand, it was recognized that time-geography is an important and powerful tool to understand the process of urbanization, urban transformation and urban development in Chinese cities from a behavioral perspective. It is necessary and a great potential for us to construct behavioral theory and the behavioral school of Chinese urban research based on the time-geographic approach in the future.

We could argue that time-geography in China has been introduced and developed against the background of international comparative research, from the earliest works by Chai with a comparative study of Chinese and Japanese cities, to learning from the USA about GIS-based space-time behavior geovisualization and geocomputation methods and the comparison between Chinese and Western cities, like Beijing and Chicago (Kwan et al. 2014; Ta et al. 2016a; Zhao et al. 2015), as well as the comparative study of cultural difference in terms of activity comparison between Beijing and Utrecht (Zhao et al. 2016). The international

Table 3.1 The history of time-geography in China

	Phase I	Phase II	Phase III	Phase IV
Period	Early 1990s to the mid-2000s	Mid-2000s to the end of 2000s	2010–2016	2017–
Main achievements	Introduction and statistical description	GIS-based 3D geovisualization and quantitative model in the framework of activity-based approach	Initiative planning practices of smart city planning Building up international research network of UCSB	Calling for theoretical development and building up international research network of time-geography
Data collection method and technology	Time use diary, activity diary (paper and pencil)	Activity diary survey and initial GPS-aided activity diary survey on the internet	GPS- and GMS-aided activity diary survey through an online survey platform	GPS-aided activity diary survey through an online survey platform Initial smartphone real-time tracking combined with environment sensors, and activity diary survey
Case study city	Lanzhou Dalian, Tianjin, Shenzhen	Beijing Guangzhou Urumqi (Xinjiang)	Beijing, Xining	Beijing Guangzhou Shanghai
Development of time-geography	Introduction of thoughts, core concepts and notation system of path, constraints and prism	Introduction of the classic empirical cases studies for planning GIS-based geovisualization of space-time path	Initial planning practices of smart travel based on individual accessibility	Introduction of concepts of project and pockets of local order Introduction of the new development of time-geography in Sweden
International cooperation	Sino-Japanese	Sino-US	Sino-US, Sino-Dutch	Sino-US, Sino-Dutch, Sino-Swedish
Major events	The first introduction of time-geography in 1997 The first Sino-Japanese comparative studies based on time-geography	Chai's (2005) review of and reflections on key issues of behavioral geography The foundation of the "Spatial Behavior and Planning research group"	2011 Chai's keynote speech in the special symposium of 2012 international workshop on "Space-time behavior and smart travel" at Peking University 2015 the establishment of the international research network of "Urban China Space-Time Behavior" (UCSB) 2015 Sino-US joint project	2014 Chai's keynote speech in the first international conference on time-geography 2015 Sino-Swedish joint project 2016 Zhang's keynote speech at the second international conference on time-geography

cooperation has led to several cutting-edge research topics, such as social seg-regation from the perspective of daily activity, dynamic geographical exposure and its social, environmental and health outcomes based on individual space-time activity, which have been put forward and will be explored in urban China. In future, greater achievements of time-geography in China are expected through more extensive and in-depth international comparison and cooperation research with Japan, the USA, the Netherlands and Sweden.

Time-geography approach to urban China's transition: some empirical studies

The time-geographic approach may be used from a micro perspective and offers a refreshing way to understand the dramatic urban social and spatial transforma-tion as well as to understand their influences on individuals' life experiences in Chinese cities driven by the dramatic market-oriented transition since 1978. New social and environmental issues arose from the market-oriented reform, which called for Chinese urban geographers to pay special attention to individual behaviors. On one hand, the choice of freedom and mobility of individual citi-zens was greatly enhanced in daily life, making people's daily activities more diversified and complicated. On the other hand, market-oriented transforma-tion and the dissolving of the *danwei* system sped up the replacement of urban land use based on land prices, promoting the urban spatial expansion as well as the rapid suburbanization. Taken together, these processes further contribute to home–work separation and longer commuting timex, a surge in travel demand, more severe traffic congestion, intensification of resource and energy consump-tion as well as environmental pollution problems and, finally, a decline in the quality of life (Chai 2014).

Changing life experiences: from danwei *to* xiaoqu

Before the economic reforms in the late 1970s, the spatial structure of Chinese cities was very different from Western cities, and the urban space was character-ized by the *danwei* system (Wang and Murie 1999). *Danwei* "is a generic term denoting the Chinese socialist work-place and the specific range of practices that it embodies" (Bray 2005). A *danwei* assumed the full responsibility of housing provision in the pre-reform period, and tended to serve as self-contained neighbor-hoods that provided housing, jobs, food distribution and other social services to their residents (Gaubatz 1999). In 1978, nearly 95% of the urban workforce were *danwei* employees (Bray 2005), and most of them lived in *danwei* compounds, which hence constituted most of their daily-life space. The market-oriented eco-nomic reform since the 1980s, with the development of the housing market in urban China has gradually removed the housing allocation responsibilities from *danwei*. The *danwei*-based spatial bond between housing and jobs (as well as ser-vices) has gradually been replaced by a market-based spatial link (Wang and Chai 2009), and the problem of spatial mismatch has become increasingly prominent.

The dynamics of the *danwei* system, and its transformation of urban development and residents' daily life, has received considerable scholarly attention, especially from those with a time-geographic perspective (e.g. Chai 1996a, 1996b, 1999). Emphasis has been put on studying the individuals' daily paths and everyday life in *danwei* compounds, and understanding the daily life from *danwei* to *xiaoqu*.

Individuals' daily paths and everyday life in traditional danwei compounds

In 1992, Chai carried out his first activity diary survey in Lanzhou in China and in Hiroshima city in Japan, and he made an international comparative study on the everyday life and on the urban structure between the two cities. The sample size was 300 couples in Lanzhou and 400 in Hiroshima respectively (Chai 1996a, 1999). It is worth remembering that Lanzhou was an inland city located in western China, and it was still mainly dominated by the *danwei* system in 1992, even though several years had passed since the market-oriented reform had been launched. Therefore, the residents' life experience in Lanzhou in 1992 could be regarded as a typical representation of that during the *danwei* period under the old redistribution system before reform.

Although much could be said about the differences between these two cities, Chai chose to compare the daily lives of both wife and husband from the same household between the two cities, using the daily path, one of the most important notations in time-geography, to demonstrate vivid pictures of real lives in different urban contexts. The purpose was to see how household members organize their daily life and carry out activities in space and time in a city context during the same 24-hour period. Figure 3.1, reveals two major differences between the two cities during the weekday. First, whether the household lived in the city center or in the suburb, it was a regular routine for Chinese residents in dual-earner households to come back home to have lunch, either only one family member coming back or both the wife and husband coming back home together at noon. On the contrary, it was less likely to happen for the couples in the Japanese city Hiroshima, where the people usually have lunch out-of-home near their workplaces (Figure 3.1 (A)–(D)). Even for the one-earner households, in which the wife didn't have a job, the working husbands in the Chinese city came back home to have lunch together with their wife and, consequently, the wives usually had to buy food or prepare food near home before their working husbands came back for lunch (Figure 3.1 (E)–(F)).

Second, the working time (start time, end time and duration of work), seemed to be equal for both working husband and working wife in dual-earner households in Lanzhou, starting at about 08:00, and ending between 17:00 and 18:00. The marked difference from Hiroshima was that working husbands had longer working hours than working wives in Hiroshima city and that working wives from households living in the suburb worked only half a day, most likely as a part-time job (Figure 3.1 (A)–(D)).

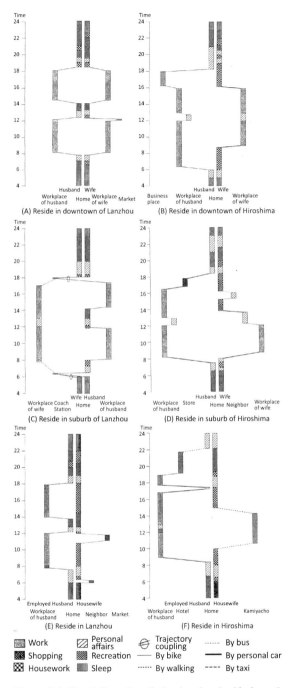

Figure 3.1 The daily paths of a husband and wife from the same household living in different contexts of Lanzhou and Hiroshima city on a weekday[5]

Modified from Chai (1999)

Furthermore, Chai explained that because most residents in Lanzhou belonged to a certain *danwei* compound and lived in housing provided by their *danwei*, and due to the job–housing adjacency under the privilege of the *danwei* welfare distribution system, the proportion of non-motorized travel was very high in Lanzhou city, with, about 50% of residents riding a bicycle to work and 35% of residents walking to work, and more than 76% of residents with a one-way commute time of no more than 15 minutes. The commuting pattern in Hiroshima city was quite different, with more than 67% of residents commuting more than 15 minutes one-way from home to work, about 44% of residents commuting by car and about 28% commuting by bus (Chai 1999). It was also because of the *danwei* system, which advocated equality between men and women with "equal pay for equal work", that there were a higher proportion of dual-earner households in Chinese cities compared to Japan.

Chai also compared the everyday life paths of husband, wife, and children from households in both a large and a small *danwei* (Figure 3.2). Within the larger *danwei* compound, almost no family members had to commute to work or school,

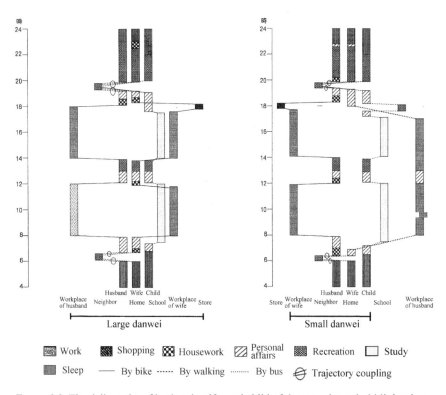

Figure 3.2 The daily paths of husband, wife, and child of the same household living in a large *danwei* (left) and a small *danwei* (right)

Modified from Chai (1999)

and their daily activity spaces largely coincided with each other within the *danwei* compound due to mixed land use and a variety of affiliated facilities provided by the *danwei*. In the smaller *danwei*, which usually could not provide as many services and facilities as the larger one did, the wife and the child had to leave home much earlier in the morning and commute longer distances to work or school (Chai 1996b, 1999).

A space-time structure map, illustrating how different subgroups allocate their daily activities in time and space during a day aggregately was developed (Figure 3.3). After a clustering analysis of the time-use structure of residents in Lanzhou, Chai identified six groups of residents with distinct time-use characteristics. In general, there existed great differences between the activity pattern of working groups (Clusters W1 to W4) and non-working groups (Clusters W5 and W6) in terms of spatial and temporal distribution. Also, for most working residents shown in Clusters W1 to W4, the space-time patterns of their working activity were very similar, which started from 08:00 and ended at about 18:00, and were located near their home (Figure 3.3).

In sum, individual paths in the household settings of Figure 3.1 show clearly the differences in everyday life in the different urban contexts of China and Japan. A richer background could be read from this kind of time-geographical graph, containing information on gender differences in activity participation in both space and time, the job–housing relationship as well as the extent of daily activity spaces. All this information was very important and could be used to establish what was unique, in terms of spatial structure and institutions, as well as cultural traditions, to each city.

Understanding daily life: from danwei *to* xiaoqu

During the past 30 years of institutional and social transformation, urban China has been undergoing a process of spatial restructuring from a *danwei*-based built environment to a new urban landscape rich in socio-spatial diversities (Wang et al. 2011). *Danwei* compounds, with a highly mixed land use, self-contained services and facilities as well as job–housing adjacency, were gradually dissolved due to the relocation of the manufacturing factories to the city outskirts, and because of the redevelopment of some old *danwei*s' residential areas driven by the force of urban land and housing marketization. At the same time, new types of commercialized residential landscapes rapidly emerged in the suburban areas, with relatively mono-functional focus on living, which further led to home–work separation and increasing travel demand, especially motorized travel. Though the *danwei* compounds and other historical traditional neighborhoods in the inner city remain important elements of major Chinese cities, new types of neighborhoods, which usually were referred to as *xiaoqu* (or literally, small district), have emerged and become equally important constituent elements of the built environment. For instance, with the urban land and housing reforms and the rise of the housing market in China, the substantial development of "commodity housing" has been witnessed, which usually takes the form of small districts or housing estates and thus forms new neighborhoods. There has been a spatial "sorting" process of *xiaoqu* based on both house price and households' housing affordability. In parallel with the development of commodity housing, social welfare housing had been introduced to provide subsidized housing

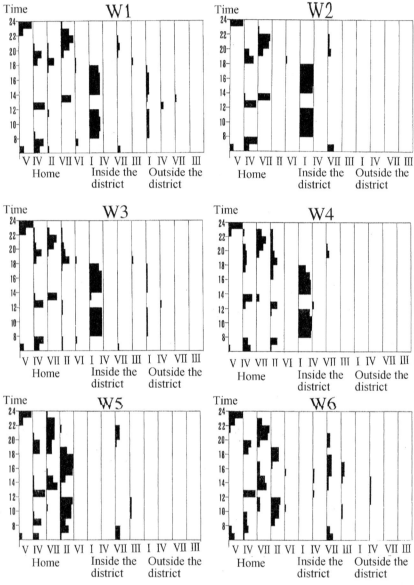

I: work; II: housework; III: shopping; IV: personal affairs; V: sleep; VI: move; VII: recreation

Figure 3.3 The spatial-temporal structure of daily activity pattern by different groups on the weekday in Lanzhou, China in 1992

Modified from Chai (1999)

for those median- to low-income households who cannot afford market-priced housing. Unfortunately, this kind of subsidized housing was mainly located in suburban areas or even on the periphery (Feng 2004).

Consequently, both the distribution of urban opportunities as well as the daily activity and travel pattern have been greatly reshaped. Researchers have attempted to explore the association between built environment and activity-travel behavior in such unique social and cultural contexts of Chinese cities especially from the perspective of daily activity of urban residents living in different neighborhoods. For instance, based on an activity diary survey carried out in 2007 in ten typical neighborhoods in Beijing with a total sample of 1,119 individuals, using the time-geographic analytical techniques (e.g. three-dimensional geovisualization analysis of the space-time path, see Figure 3.4), Zhao and Chai (2013) found that *danwei* residents have less daily travel duration and higher proportion of non-motorized trips than those who live in commodity housing, but people living in affordable housing endure the longest travel time. They argue that although China's transition is currently gradual, the *danwei* system may continue to play significant roles in residents' daily life.

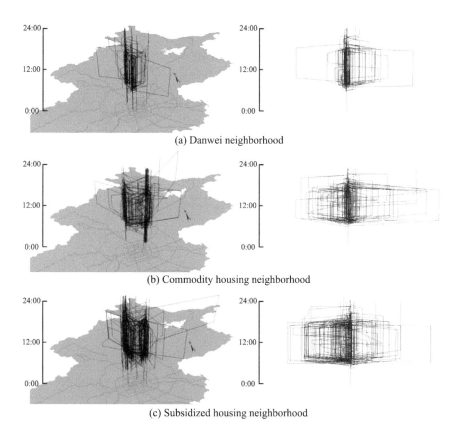

(a) Danwei neighborhood

(b) Commodity housing neighborhood

(c) Subsidized housing neighborhood

Figure 3.4 Space-time path of residents in different types of neighborhoods in Beijing
Modified from Zhao (2016)

The spatial and temporal distribution of working and non-work activities of residents living in different neighborhoods gives rise to obvious diversities and variations between the various types of neighborhoods. For the residents in traditional *danwei* neighborhoods, work activities were much more concentrated around their home locations on workdays. For the residents in historical residential areas, named *hutong* district, as well as those who live in the transformed *danwei* neighborhoods where either the factories or the residential area were relocated or redeveloped under the process of marketization, the work activities were dispersed all over the city but mainly within 10–15 km away from their home location. In contrast, for the residents in newly built commercial housing neighborhoods, and in the subsidized housing neighborhoods, which were usually called *xiaoqu*, work activities were much more scattered with a larger spatial scales even more than 20 km far away from their home locations. Besides, the out-of-home non-work activity on weekends, the situation was quite different. Residents in *danwei* neighborhoods as well as in the commercial housing neighborhoods, had a broader range of out-of-home activities. In contrast, a large percentage of residents in *hutong* neighborhoods and subsidized housing neighborhoods had a more concentrated activity space near their home location (Zhang et al., 2014). Therefore, significant differences were found in the usage of time and space between residents inside and outside the so-called privileged enclaves. The activity spaces are found to vary significantly in terms of extensity, intensity, and exclusivity. The fragmentation of urban space is the result not only of residential segregation, but also of how different social groups spend their time and use urban space (Wang et.al. 2012).

In the 30-year process of changed urban organization from *danwei* to *xiaoqu*, not only the built environment but also the residents' daily activities and travels have undergone profound transformations with a trend towards diversity and

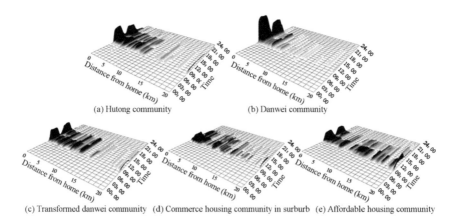

(a) Hutong community   (b) Danwei community

(c) Transformed danwei community   (d) Commerce housing community in surburb   (e) Affordable housing community

Figure 3.5 The space-time distribution of work activity of residents living in different types of neighborhoods in Beijing, 2007

Modified from Zhang et al. (2014)

complexity, which further led to more social and environmental outcomes at the macro level for the whole city. Other studies also provided evidence that *danwei* neighborhoods were associated with shorter commute distances (Liu et al. 2009; Chai et al. 2010), more non-motorized travel (Wang and Chai 2009; Wang et al. 2011), and higher probabilities of low travel-carbon emissions (Liu et al. 2014). Moreover, the diversity of neighborhood types in Chinese cities has been found to be notably related to residents' space-time constraints. As indicated in the research of Shen et al. (2015), for residents in *danwei* communities and affordable housing communities, space–time fixity of their activities was not so sensitive to the physical built environment, which suggests that the structure of their space–time constraints was influenced more by institutional factors than by spatial factors, since *danwei* communities are part of the socialist planned economy, and affordable housing communities are controlled by the government and highly related to housing policy.

Inspired by time-geography, especially the key methodological developments such as collection of high-quality space-time data and development of GIS-based analytic techniques, tremendous progress had been made in the field of space-time behavior research in the urban context of China. Empirical themes have expanded from the space-time patterns of daily activities and travel behavior to more urban issues related to social equity, public health and environmental sustainability (e.g., Lan and Feng 2010; Zhang and Chai 2011; Ma et al. 2011; Wang et al. 2012). Notably, urban geographers in China have interpreted micro-level patterns of space-time behavior within the context of macro-level institutional transformation and spatial restructuring in Chinese cities. The major conclusion from the above-mentioned empirical research is that, in the transition from *danwei* to *xiaoqu*, there are not only changes in the built environment, but also institutional transformations due to the market-oriented reform, and both of them have important influences on individuals' daily life. In transitional Chinese cities, on one hand, we have seen increasing market forces which shape individuals' spatial choice and everyday life, and, on the other hand, we have found that *danwei*, as a former redistribution institution as well as a basic spatial territory of social governance, still deeply affecting the residents' space-time behavior, lifestyle and even quality of life (Zhang 2015).

Combining time-geography with life course theory: life path and institutional transition

The seeds of what later became the time-geographic approach originates from Hägerstrand's early studies of population migration in the Asby area in Sweden. By then his research concerned spatial and temporal processes of residential migration on a lifetime scale. Yet, more recently it was mainly recognized as an important methodological foundation of daily activity analysis, and more empirical studies applied its core concepts and notation system to short-term behavior at the daily level. Considering the uniqueness of the institutional transition of Chinese cities, we constructed a new conceptual framework which integrates time-geography with the life course theory for a better understanding of long-term space-time behaviors (Chai et al. 2013).

Life course theory, a sociological approach established in the 1960s, focused on analyzing people's lives within structural, social and cultural contexts, with the emphasis on the relationship between individual life course and social change. It tried to connect the individual meaning of life with social significance by describing the structure and sequence of individual trajectories and transitions during the whole life. We argued that both time-geography and life course theory focus on the individual's overall life changes from a process-oriented perspective, while emphasizing time and space. The difference between the two is that the life path in time-geography values the spatial change of place, while the life trajectory in life course theory focuses more on the change of the mental or social state. Both of them attach importance to the role of constraints, and time-geography pays more attention to the materialist basis of physical entities. In contrast, life course theory emphasizes the influence of social structure and social relations, and the analysis of physical space is not enough. Both approaches involve long-term and short-term relationships, as well as the dynamic of the past, present and the future. Therefore, we further argued that the life course approach could supplement time-geography with an in-depth understanding of the social context of an individual's long-term space-time behavior process and of the time accumulation effect of past and future behavior. Therefore, we re-conceptualize the notation of life path by integrating individuals' life events in different life spheres, such as residential migration, change of employment, marriage and children, and so on, against the macro background of social change. We also enrich the life path by adding individuals' subjective factors, such as emotions and feelings as well as other social and psychological factors. In the following section, we take the behavior of residential relocation as an example to illustrate the contribution of combining these two approaches to study long-term spatial behavior.

First, time-geography emphasizes individual indivisibility, and life course theory posits that individual life events and social roles are interrelated. The life path does not focus on the change of residence, but rather depicts the individual's occupational history, family changing history, etc. Therefore, in order to reflect the indivisibility of the individual, instead of depicting the individual residential movement path and the individual career change paths respectively, we integrated these two paths into one life path (Figure 3.6). With this multi-dimensional life path, we could analyze the different types of personal life events before and after and recognize how they exert influence on each other, as well as the function of the personal effort to make changes in the path of life. In addition, by introducing the concept of "turning points" into life course theory, we emphasize the great impact of the critical point on individuals' long-term behavior by marking the life events of the individual life path. The person in Figure 3.6 is described by two life paths: to the left is his migration path, and on the right is his career change path. Although these two paths can be analyzed independently, we can also combine the person's residential movement path and his career change path according to the order of the life time, and do some comparative analysis. From the residential movement path of the person in

Figure 3.6, it is shown that he has moved three times in his life. His first residence was the home where he lived together with his parents as a child and adolescent. Combined with the career changing path on the right, we could find that he went to school during that period. Later, he moved out of his parents' home to an apartment provided by his employer (the work unit, a *danwei*) where he worked. This was the first residential move in his life. After that he experienced several important life events: marriage, the birth of a child, and changes of jobs, but still belonged to the same *danwei*. By the end of the 1970s he moved to larger housing built by the *danwei*, where he has lived since then. He experienced other important life events such as his child growing up and leaving home to start a new family, he got retired and his wife passed away. Based on this multi-path picture of both residential history and career history, we could add the subjective dimensions of his life path, such as subjective satisfaction, feelings, emotions and intentions for changing living conditions and so on, which could deepen our understanding of how the institutional transition influenced the individual life path.

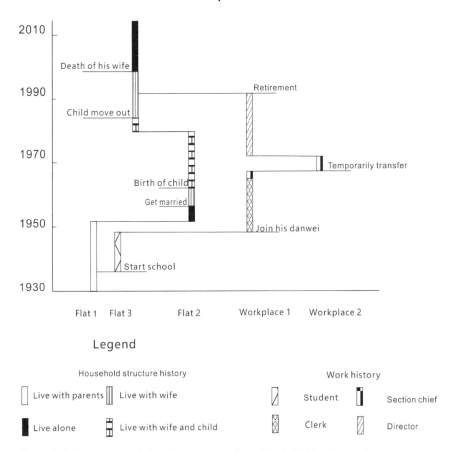

Figure 3.6 Re-conceptualizing the representation of the individual life path
Modified from Chai et al. (2013)

Based on the combination of life course theory and time-geography, there was a qualitative case study of the impacts of residential mobility on the daily activity pattern against the background of suburbanization in Beijing in 2015 (Figure 3.7) (Chai and Feng 2016). On the left in the figure is a person's migration path and his career change path, on the right are his daily paths in three different stages: after getting married, before and after the first child was born. After getting married, he moved to the community near his wife's workplace, which allowed his recreation activities to become more home-centered. When his first child was born, his parents came to Beijing to help the couple take care of the child. The conflict between increasing family size and small living space made the couple decide to move to a new house. They considered the housing area and price, the change of wife's workplace and the affordability of the family, and finally they chose a house close to the wife's workplace but far away from his, which completely changed his daily activity. His daily path began to approach that of typical long-distance commuters. We could see that changes in the family structure will lead to changes in housing demand, and a family's relocation decision often is the result of consultation

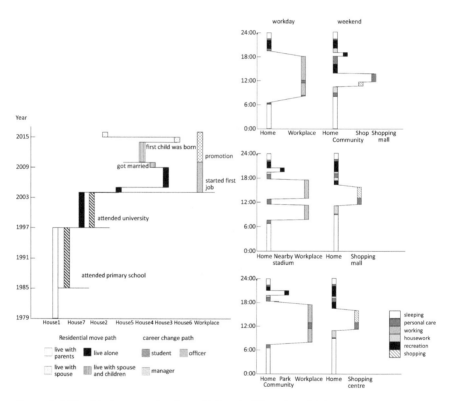

Figure 3.7 Empirical case study of combining the life path and daily path and focusing on the important impacts of life events, such as getting married and birth of child on daily life experiences

Modified from Chai and Feng (2016)

and compromise among family members. This relocation will change the relation between residence and employment of the family member, which directly affects their daily activities and potential opportunities.

Daily life space and lifestyle of different social groups under the urban transition

In recent decades, increasing income disparity and socio-spatial differentiation have been witnessed in urban China along with waves of economic transition. Particularly, since the housing system reform in the 1990s, socio-spatial segregation has become increasingly prominent in mega-cities in China, and has received considerable scholarly attention. However, studies on socio-spatial segregation from the perspective of space-time behavior was rarely carried out until the introduction of time-geography to Chinese researchers. Since then, a series of empirical studies based on time-geography have been conducted, which considerably contribute to the existing literature on socio-spatial differentiation in urban China.

These studies argue that people living in the same residential area would not necessarily experience the same level of socio-spatial differentiation. Rather, socio-spatial differentiation may take place in daily activity spaces other than the residential area as everyday life unfolds. In this regard, Wang et al. (2012) suggest four dimensions along which activity space may be measured, namely, extensity, intensity, diversity and exclusivity, in order to quantify how different groups make use of urban space as a result of socio-spatial differentiation. Shen and Chai (forthcoming) propose a research framework of "people–activity space–social space" to investigate how activity spaces vary across different social groups as well as to identify disadvantaged groups through the manifestations of activity spaces. Tan (2017) expands the existing residential exposure indices to examine different ethnic groups' mutual exposure in their activity spaces.

In terms of empirical studies, three major topics could be identified, namely, income level, gender and ethnic group. Studies concerning these topics prove that activity space is a helpful perspective for achieving a deeper understanding of socio-spatial differentiation through focusing on the dynamic nature of everyday life.

Income level

Researchers have made several attempts to reveal the variations of space-time behavior by the disparity of income level, which provides meaningful insights into social equality issues. Based on an activity diary survey, and aided by GIS-based geovisualization technology, activity spaces and people-based accessibility were calculated in order to discover the socio-spatial differentiation of space-time behavior. For instance, based on the activity diary data of Guangzhou collected in 2007, Zhou and her groups revealed significant differences in the space-time behavior between low- and high-income residents, the low-income residents tending to have restricted activity spaces and longer out-of-home activity duration (e.g. Zhou and Deng 2010; Zhou et al. 2015). Zhang and Chai (2011) investigated the spatio-temporal patterns of daily activities of the low-income residents in Beijing in terms of activity time-use, spatio-temporal distribution of working and non-working activities, and daily mobility, and also found similar results. Liu et al. (2005) analyzed the daily paths of the poverty group in Nanjing and pointed out that greater attention should be paid to neighborhood-level planning for the urban poverty group due to the fact that their daily activities were more concentrated around their home locations. Ta and Chai (2017) shed light on the potential activity spaces of low-income residents, and tried to reveal the accessibility inequality among different income groups, providing new insights into income-based socio-spatial differentiation in urban China.

Gender

As described earlier, and different from Western cities, there has been a very high female labor force participation rate in urban China. The dual-earner household is a very common phenomenon in China. Also, the male household members took on more domestic tasks, and gender difference in domestic labor division was significantly less in urban China than in Western and Japanese cities (Sheng et al. 1992). Furthermore, there exist distinctive social and institutional factors which mediate the negotiation process of domestic labor division in dual-earner households in transitional urban China. First, due to the legacy of the *danwei* system, which might still facilitate working women to reconcile employment and caregiving tasks by providing them with both institutional employment support and childcare services (Stockman 1994). Second, it is a cultural norm in urban China that married couples could get support for caregiving tasks from their kin networks, especially from grandparents. Therefore, gender difference in everyday activity is one of the most interesting topics on which great attention has been focused.

Among several studies based on the early round of activity diary surveys by the end of the 1990s, Cao and Chai (2007) focused on time-use patterns and found that women carry more household responsibilities than men and that men spend more time on work and leisure activities. Zhang and Chai (2008) further revealed the gender-based division of domestic labor in Chinese urban households by comparing space-time activity patterns between husbands and wives from dual-worker households in Dalian, Tianjin and Shenzhen. They argued that "men are dominant in out-of-home activities, but women dominate in-home activities" as wives had lower peaks in work activities whereas they had higher peaks in household activities (Zhang and Chai 2008: 1254).

During the decades after the abolition of the *danwei* system, an empirical study based on the activity diary survey in Beijing in 2007 revealed that the legacy of the *danwei* still plays an important role in domestic labor division in dual-earner households and further on gender difference in daily activity and travel. By exploring the determinants of domestic labor division based on 240 dual-earner households in Beijing, we found that households living in suburban neighborhoods including commercial housing and subsidized housing neighborhoods were more likely to divide household tasks in a non-traditional way than were households in *danwei* neighborhoods. This indirectly supports the hypothesis that the legacy of *danwei* would be helpful for women to combine paid work and household tasks, which in turn increases the likelihood of a traditional division of domestic labor between partners. Further, male partners' sharing of domestic labor significantly shaped the activity pattern of working women on weekdays, and the division of domestic labor would mediate the effect of gender role on an individual's space-time activity pattern (Zhang 2015).

Another recent study further points out that both household structure and gender significantly impact the space-time constraints measured by the temporal and spatial fixity of each activity on a daily basis. Both men and women in extended family households where the working couple could get childcare provided by grandparents, perceive less temporal fixity (but not spatial fixity) than men and women in nuclear family households. The impact of household structure on activity participation and space-time constraints is greater for women than for men and greater on weekdays than on weekends (Ta et al. 2016b).

In all, different institutional, spatial and social contexts at neighborhood and household levels between Chinese cities and Western cities provide great opportunity for examining gender roles and their implications for daily life. Time-geography also provides a unique perspective as well as a sensitive tool to explore the complexity and diversity of this time. In future, more empirical studies armed with time-geography are expected.

Ethnic groups

Ethnic differences in space-time behavior have also been investigated by Chinese researchers. For instance, Zheng and his colleagues explored the differences in space-time behavior (e.g. shopping activities) between Uygur and Han ethnic groups across weekdays and weekends in Urumqi, China (Zheng et al. 2011; Zheng et al. 2009). Using two-day activity diaries, Tan et al. (2017a, 2017b) applied analytical techniques based on time-geography (e.g. space-time path, space-time trend surface, activity spaces) to identify differences in space-time behavior and activity-space segregation between Hui and Han ethnic groups in Xining city. They conclude that participation in spatially and temporally fixed daily religious activities has led to an independent and significant influence of ethnicity on space-time behavior when compared with the Han majorities in Xining. Moreover, based on the same data set, Tan et al. (2017c, 2017d) explore the impacts of geographic context on ethnic segregation, which sheds new light on ethnic-based differentiation in urban China from the perspective of time-geography.

Smart community planning research and practices in urban China

In addition to empirical research, Chinese scholars have tried to apply the space-time behavior approach in urban planning practices, including transportation planning and management, smart city master planning, tourism planning, and infrastructure site planning. The space-time behavior approach has been viewed as a human-oriented approach to urban and transportation planning, as urban planning is transforming from serving the purpose of economic growth to promoting livable and sustainable cities (Liu et al. 2009; Chai et al. 2010). In particular, urban scholars and planners have begun to realize the importance of serving the diverse needs of urban residents in transitional urban China. To meet this goal, the space-time behavior approach offers an effective way for understanding individuals' spatio-temporal demands in order to adjust the urban spatial planning and spatial structure to improve the quality of urban life in China.

For instance, by performing geovisualizations and simulations of space–time behavior, researchers have attempted to map and predict visitor flows and activity patterns in tourist sites and public spaces (e.g. the Summer Palace in Beijing, Shanghai Expo, see Zhu and Wang 2008; Wang et al. 2009; Huang and Ma 2011; Huang and Wu 2012). By integrating activity-travel data (derived from the mobile and taxi GPS data), land-use data, population and economic census data, researchers developed a diagnostic index of urban signs that serves to detect the spatial-temporal patterns of human activities in Shanghai, which has huge potential for implementation in urban grid management, pressure warning and other needs of urban governance (Chai et al. 2018). Residents' daily life circles[6] that capture the access to urban facilities are also identified based on the modeling of activity spaces using GPS data (Sun et al. 2016; Sun and Chai 2017).

Smart travel planning: facilitating travel decision making based on activity-travel analysis

An initiative to apply space-time behavior research to smart travel planning has been carried out in Beijing based on the theoretical framework of time-geography since 2012 (Chai and Chen 2018). The framework of the project has three steps. First, by using location-aware technology, multi-source activity-travel data were collected and integrated with the built environment data. Second, residents' activity-travel patterns and determinants of travel decision making were revealed based on the time-geographic analysis of data. Third, by developing a mobile-based smart travel planning system, the results of the analysis were further transferred into real-time information and delivered to residents, which served to promote efficiency in travel decision making. Specifically, the smart travel planning system is used on a portal website as well as a mobile app. Several types of activity-travel information, including maps of walkability services, travel heat maps and real-time traffic conditions, were integrated into the system. Users may log on to the system, choose their community, and read the dynamic information published on the system, which provides beneficial support to their travel decisions. Overall, this project has made an attempt to apply individual activity-travel data and time-geographic analysis in a mobile-based travel planning service, aiming to promote efficiency in travel decision making as well as to relieve traffic problems at the aggregate level.

Social sensing of the city: construction and calculation of a diagnostic index of urban signs

The recent tidal wave of individual-level big geospatial data, which provides a brand-new opportunity for social sensing, has caught the attention of researchers and planners investigating individual mobility patterns (Yue et al. 2014; Liu et al. 2015), with meaningful implications for urban management and early warning systems of emergencies. Using a time-geographic framework, researchers developed a diagnostic index of urban signs that serves to detect the spatial-temporal patterns of human activities in Shanghai (Chai et al. 2018). Urban signs characterize the state of development and operation of a city in terms of built environment, transportation and activity-travel system, which could be compared to a health examination of urban development and operation. Based on activity-travel data (derived from mobile and taxi GPS data), land-use data, population and economic census data, a system of reliable and practical urban diagnostic indices was developed. Specifically, twelve spatial-temporal scales are acquired through the intersection of four levels of spatial units (municipal Shanghai, district, Jiedao, census tract) and three levels of temporal scales (annual, daily, real-time), which highlight the dynamic characteristics of the activity system and flows in time and space. Through visual and real-time analysis and evaluation, the diagnostic index of urban signs has huge potential for implementation in urban grid management, pressure warning and other needs of urban governance.

Life circle planning based on space-time behavioral analysis

Based on the perspective of time-geography, researchers have attempted to identify daily life circles based on residents' routine activities (Chai et al. 2015; Liu and Chai 2015; Sun et al. 2016; Sun and Chai 2017). Specifically, based on the analysis of residents' space-time behavior with regard to different neighborhoods, general models of daily life circle are proposed and applied in community planning. For instance, community life circles are defined by the basic maintenance or leisure activities near residential neighborhoods (e.g. shopping, physical activities and leisure activities around the neighborhoods). The overlaps of neighboring community life circles (with shared facilities) constitute the basic life circles. Besides, commuting life circles are defined as the commuting-related activity spaces (with workplaces being the anchor points). Both working activities and some non-working activities on the commuting trains (e.g. dining and shopping) are involved to delineate commuting life circles. Moreover, expanded life circles are defined by some of the occasional activities, e.g., leisure activities on weekends that are distant from home. They are structured by the anchor points of several discretionary activities.

Overall, these studies expanded the concept of community life space to the scale of the whole city, and put forward the idea of urban life space planning, which refers to the reconstruction of life space by reorganizing community activity, commuting activity, leisure activity, shopping activity and so on, based on the notion of urban life space, in order to explore the ideal space pattern and to pursue the optimal scheme by which the urban space and the residents' behavior can match. These studies provide meaningful implications for understanding and planning Chinese cities from the perspective of time-geography. They not only identified the existing problems of community planning, but also made an attempt to reconstruct the current planning paradigm, and shed light on the people-oriented planning in China in the context of social transformation.

Application in tourism planning

Tourism planning from the time-geographic perspective focuses on the experiences of tourists, and addresses the individualized and preference-based tourism services and facilities. Also, the empirical studies based on time-geography provide meaningful implications for tourism market analysis and strategic planning. For instance, Zhao et al. (2017) explored the impact of tourists' participation on behavior, using the framework of space-time constraints, based on an empirical survey in Ocean Park Hong Kong (OPHK) that combined GPS (Global Positioning System) tracking and a paper-and-pencil questionnaire. They revealed that the fixity of shows or performances caused strong constraints on tourists' previous activities and resulted in small activity spaces, and thereby proposed that the planning of shows or performances should take into account tourists' space-time accessibility. By investigating the tourists' behavior in the Summer Palace of Beijing, Huang (2009) identified the potential of time-geography in dynamic management of tourist attractions, route optimization and time planning. Wang and his colleagues (Wang et al. 2015; Wang and Ma 2009) modeled the spatio-temporal patterns of tourists in mega-exhibitions (e.g. Shanghai Expo) in terms of visitors' flow, dining and rest demand, visiting time and routes, etc., which provides implications for exhibition management and infrastructure site planning. The modeling results show great similarity with the real situations as indicated in a follow-up study, which testified to the validity of the methodology.

Conclusion and discussion

The history of time-geography in China, and the overview of how the time-geographical approach is applied, both for empirical studies and planning practices over the last three decades, provides the basis for the argument that time-geographic research offers a new understanding of the complexity and diversity of human behavior patterns during the large-scale spatial and institutional transition in Chinese cities. From the West, Chinese research learnt to improve data quality and analytic tools, which has facilitated GIS-based geovisualization of space-time activity patterns in Chinese cities and improved the understanding of space-time behavior decisions within the constraints set by the built environment.

Nonetheless, we also recognize that while the techniques and tools used in the visualization of individual path and space-time activity patterns in China have caught up with those in the West, most empirical studies using the time-geographic approach or "space-time" behavior research have not seriously looked into the roles or principles behind individuals' overt behavior, or space-time trajectories. In other words, the studies on space-time constraints were inadequate compared to the studies on space-time pattern. In most cases, we collected the individual-level data, and depicted the individual space-time trajectory of a certain sub-population together and expected to find a pattern by doing so. Yet, coming back to the starting point of time-geography, we should look more carefully at the micro settings where the individual participates in space and time, in order to find both the constraints and opportunities for him given the specific circumstances to fulfill certain projects which were formed earlier. Accordingly, it is not the individual per se, but the individual together with other associated people or things in certain time-space settings that should be our study object. We firmly assert that there is much we should do in that manner that could provide a deeper understanding of micro processes and direct interactions between space and behavior.

In addition, as a major goal for Chinese geographers insisting on a behavioral perspective, we aim to build up a theory of urban transition from the perspective of the interaction between space and time, which might be quite different from the institutional school (Chai et al. 2016, 2017). We firmly believe that time-geography would provide a strong basis for building up that theory. The most important thing that should be given more attention, and which up till now has not been clear enough, is that "space-time behavior" differs fundamentally from "spatial behavior". The space-time is not equal to space plus time, but means the complex and dynamic context in which individuals behave. Also, an individual-oriented approach does not mean that we should depict the movement of individuals one by one, on the contrary, we should pay much attention to how the individual intervolves with other individuals, with groups or institutions together. By doing so, we need to carry out empirical studies using other key concepts of time-geography, e.g. project and pocket of local order, which we had not fully recognized before, to answer how and why questions of the interactions between individual space-time experiences and the urban transformation process. By doing so, we could enhance our contribution to the future theoretical development of the time-geographic approach.

Acknowledgements

This work was supported by the National Natural Science Foundation of China (grant no. 41611130051, grant no. 41771185, 41571144, 41529101). The authors would like to thank Professor Kajsa Ellegård for her kindly invitation for us to attend the first and second international conferences on time-geography and to give keynote speeches. This chapter partly originates from these speeches. Also, we would like to express our acknowledgements to the following core members of the time-geography research group in Peking University, including Dr Zhilin Liu who is an associate professor of the Public Management School of Tsinghua University, Dr Wenjia Zhang who is an assistant professor of Peking University (Shenzhen), Dr Ying Zhao who is an associate professor of Sun Yat-sen University, Dr Yue Shen who is an associate professor of East China Normal University, Dr Na Ta who is an associate professor of East China Normal University, Dr Jing Ma who is an assistant professor of Beijing Normal University, and Dr Zuopeng Xiao who is assistant professor of Harbin Institute of Technology (Shenzhen).

Notes

1 The research network of "Spatial Behavior and Planning" was formed in 2005. Annual conferences have been held by this research network around China since 2007. And up to 2017, there have been thirteen annual conferences, including the 2007 annual conference in Hong Kong, the 2008 annual conference in Changchun, the 2009 conferences in Shanghai and Beijing, the 2010 conference in Beijing, the 2011 conference in Guangzhou, the 2012 conferences in Beijing and Hong Kong, the 2013 conference in Nanjing, the 2014 conference in Shanghai, the 2015 conference in Guangzhou, the 2016 conference in Beijing, and the 2017 conference in Hangzhou. The total number of conference participants has increased from less than thirty to over three hundred recently.
2 The *danwei* system refers to the institutional arrangements under the planned economy in urban China, concerning resource allocation, employment and social welfare and with unique spatial and social representations. The *danwei* (or work unit) compound was the basic spatial and social unit of urban China in the pre-reform period. *Danwei* "is a generic term denoting the Chinese socialist work-place and the specific range of practices that it embodies" (Bray 2005). A *danwei* assumed the full responsibility of housing provision in the pre-reform period, and tended to serve as a self-contained neighborhood that provided housing, jobs, food distribution and other social services to its residents (Gaubatz 1999).
3 Yanwei Chai studied abroad at Hiroshima University in Japan from 1988 to 1994. Since the 1980s, Japanese geographers have taken a tremendous interest in the time-geography approach. They introduced the time-geography approach into the Japanese geographical society and applied it in empirical studies of everyday life and life space of rural residents in Japan, married women and a variety of restrictive factors affecting their entry into the labor market as well as life and activity space in urban contexts (Arai et al. 1989, 1996). At this stage, Chai was a doctoral student in Japan. He paid attention to those studies and considered that time-geography was also applicable to the Chinese context, especially in terms of the urban problems accompanying the rapid economic development in China. He carried out the first activity diary survey in Lanzhou City, the capital city of Gansu Province in western China during 1990–1992.
4 "GIS methods in time-geographic research: Geocomputation and geovisualization of human activity patterns"; and the other was from Shaw in 2009, "A GIS-based time-geographic approach of studying individual activities and interactions in a hybrid physical–virtual space".
5 Note that it is not a mistake that in Figure 3.1 (C) the wife is to the left and the husband is to the right, while they are placed in opposite positions in all the other parts of the figure.
6 Daily life circle refers to the everyday activity space of Chinese urban residents which takes their neighborhood as a core and, approximately within the spatial scope of 15 minutes' walking, non-work activities. Before the market-oriented reform, especially

under the *danwei* system, a resident's daily life circle was basically within his *danwei* compound. In contrast, during the transitional period, a resident's daily life circle in most cases was around his *xiaoqu*.

References

Arai, Y., Kawaguchi, T, Okamoto, K., and Kamiya, H. 1989. *The Space of Life the Time of City.* Tokyo: Kokon (in Japanese).

Arai, Y., Okamoto, K., Kamiya, H., and Kawaguchi, T. 1996. *The Space and Time of City: The Time-Geography of Daily Activity.* Tokyo: Kokon (in Japanese).

Bray, D. 2005. *Social Space and Governance in Urban China: The Danwei System from Origins to Reform.* Redwood City: Stanford University Press.

Cao, X., and Chai, Y. 2007. Gender role-based differences in time allocation: Case study of Shenzhen, China. *Transportation Research Record Journal of the Transportation Research Board,* 2014, 58–66.

Chai, H., and Feng, J. 2016. Behavior space of suburban residents in Beijing based on family life course. *Progress in Geography,* 35(12), 1506–1516 (in Chinese).

Chai, Y. 1996a. The internal structure of a city in Chinese arid areas: A case study of Lanzhou, Gansu Province. *Chinese Journal of Arid Land Research,* 9(3), 169–180.

Chai, Y. 1996b. *Danwei* -based Chinese cities' internal life-space structure: A case study of Lanzhou City. *Geographical Research,* 15(1), 30–38 (in Chinese).

Chai, Y. 1998. Time-geography: Its origin, key concepts and applications. *Scientia Geographica Sinica,* 18(1), 65–72 (in Chinese).

Chai, Y. 1999. *A Comparative Study on the Urban Spatial Structure between Chinese Cities and Japanese Cities.* Beijing: Peking University Press (in Chinese).

Chai, Y. 2005. Methodological problems in behavioral geography study. *Areal Research & Development,* 24(2), 1–5 (in Chinese).

Chai, Y. 2013. Space-time behavior research in China: Recent development and future prospect. *Annals of the Association of American Geographers,* 103(5):1093–1099.

Chai, Y. 2014. From socialist *danwei* to new *danwei*: A daily-life-based framework for sustainable development in urban China. *Asian Geographer,* 31(2), 183–190.

Chai, Y., and Chen, Z. 2018. Towards mobility turn in urban planning: Smart travel planning based on space-time behavior in Beijing, China. In Shen, Z., and Li, M. (Eds.). *Big Data Support of Urban Planning and Management* (pp. 319–337). Cambridge: Springer.

Chai, Y., and Gong, H. 2000. The time-geography research focuses on life quality. *Bulletin of the Chinese Academy of Sciences,* 2000(6), 417–420.

Chai, Y., and Gong, H. 2001. Time-geographical approach to the study of urban society. *Journal of Peking University: Humanities and Social Sciences,* 38(5), 17–24 (in Chinese).

Chai, Y., and Liu, X. 2002. A time geographical framework and prospect of urban ageing study, *Areal Research and Development,* 21(3), 55–59 (in Chinese).

Chai, Y., and Wang, E. 1997. Basic concepts and notation of time-geography. *Economic Geography,* 17(3), 55–61 (in Chinese).

Chai, Y., and Zhang, Y. 2011. Hägerstrand and "What about people in regional sciences?" In Chai, Y., and Wyckoff, B. (Eds.). *The Interpretation of Classical Thoughts in Geography.* Beijing: Commercial Press (in Chinese).

Chai, Y., and Zhang, Y. 2012. Time-geography and its application in urban geography. In Chai, Y. et al. (Eds.). *Thoughts and Methods in Urban Geography* (pp. 201–213). Beijing: Science Press (in Chinese).

Chai, Y., and Zhao, Y. 2009. Recent development in time-geography. *Scientia Geographica Sinica*, 29(4), 593–600 (in Chinese).

Chai, Y., Li, Z., Liu, Z., and Shi, Z. 2000. A review of time-geography. *Human Geography*, 15(6), 58–63 (in Chinese).

Chai, Y., Liu, Z., Li, Z., Gong, H., Shi, Z., and Wu, Z. 2002. *The Temporal and Spatial Structure of Chinese Cities*. Beijing: Peking University Press (in Chinese).

Chai, Y., Zhang, W., Zhang, Y., et al. 2009. The production and quality management of disaggregated space-time data of individuals' behaviors. *Human Geography*, 24(6), 1–9 (in Chinese).

Chai, Y., Li, C., Liu, X., and Cao, L. 2010. *The Activity Space of the Elderly in Urban China*. Beijing: Science Press (in Chinese).

Chai, Y., Zhao, Y., and Zhang, Y. 2010. Time-geography and its application in urban planning. *Urban Planning International*, 25(6), 3–9 (in Chinese).

Chai, Y., Kwan, M., and Shaw, S. 2010. Time-geography and urban planning: An introduction. *Urban Planning International*, 25(6), 1–2 (in Chinese).

Chai, Y., Ta, N., and Zhang, Y. 2013. Rethinking time-geography in long-term space-time behavior study: Integrating with life course theory. *Human Geography*, 2013(2), 1–6 (in Chinese).

Chai, Y., Chen, Z., Liu, Y., and Ma, X. 2014. Space-time behavior survey for smart travel planning in Beijing, China. In Rasouli, S., and Timmermans, H. (Eds.). *Mobile Technologies for Activity-Travel Data Collection and Analysis* (pp. 79–89). Hershey, PA: Information Science Reference IGI.

Chai, Y., Liu, Z., Zhang, Y., Ta, N. 2015. Space-time behavior research and application in China. In Kwan, M.-P., Richardson, D., Wang, D., and Zhou, C. (Eds.). *Space-Time Integration in Geography and GIScience* (pp. 21–38). Dordrecht: Springer.

Chai, Y., Zhang, X., and Sun, D. 2015. A study on life circle planning based on space time behavioral analysis: A case study of Beijing. *Urban Planning Forum*, 3, 61–69 (in Chinese).

Chai, Y., Ellegård, K., and Zhang, Y. 2016. Introduction to the special issue on "new" time-geography. *Human Geography*, 2016(5), 17–18 (in Chinese).

Chai, Y., Ta, N., and Ma, J. 2016. The socio-spatial dimension of behavior analysis: Frontiers and progress in Chinese behavioral geography. *Journal of Geographical Sciences*, 26(8), 1243–1260.

Chai, Y., Tan, Y., Shen, Y., and Kwan, M.-P. 2017. Space-behavior interaction theory: Basic thinking of general construction. *Acta Geographica Sinica*, 36 (10): 1959–1970.

Chai, Y., Liu, B., Liu, Y., et al. 2018. Construction and calculation of diagnostic index of urban signs based on multi-source big data: Case of Shanghai. *Scientia Geographica Sinica*, online (in Chinese).

Chen, J., Shaw, S.-L., Yu, H., et al. 2011. Exploratory data analysis of activity diary data: A space–time GIS approach. *Journal of Transport Geography*, 19(3):394–404.

Chen, J., Lu, F., Zhai, H., and Shaw, S.-L. 2015. Making place recommendations: An individual accessibility measure to urban opportunities in space and time. *Acta Geographica Sinica*, 2015(6), 931–940 (in Chinese).

Chen, J., Shaw, S.-L., and Feng L. U. 2016. A space-time GIS approach for human behavior studies. *Journal of Geo-Information Science*, 18(12), 1583–1587 (in Chinese).

Dai, D., Zhou, C., and Ye, C. 2016. Spatial-temporal characteristics and factors influencing commuting activities of middle-class residents in Guangzhou City, China. *Chinese Geographical Science*, 26(3), 410–428 (in Chinese).

Ellegård, K., Liu, B., Zhang, Y., and Chai, Y. 2016a. Concept of project in time-geography and its empirical case studies. *Human Geography*, 31(5): 32–38 (in Chinese).

Ellegård, K., Zhang, X., Zhang, Y., and Chai, Y. 2016b. Pockets of local order and its application in human activity research. *Human Geography*, 31(5): 25–31 (in Chinese).

Ellegård, K., Zhang, Y., Jiang, C., and Chai, Y. 2016c. Visualization and applications of daily activities in the complex context. *Human Geography*, 31(5): 39–46 (in Chinese).

Feng, J., 2004. *Spatial Restructuring of Chinese Cities in the Transition Period*. Beijing: Science Press (in Chinese).

Gaubatz, P. 1999. China's urban transformation: Patterns and processes of morphological change in Beijing, Shanghai and Guangzhou. *Urban Studies*, 36(9), 1495–1521.

Gu, J., Zhou, S., Yan, Z., and Deng, L. 2012. Formation mechanism of traffic congestion in view of spatio-temporal agglomeration of residents' daily activities: A case study of Guangzhou. *Scientia Geographica Sinica*, 32(8), 921–927 (in Chinese).

Gu, J., Zhou, S., and Yan, X. 2013. The space-time paths of residential mobility in Guangzhou from a perspective of life course. *Geographical Research*, 32(1):157–165 (in Chinese).

Gu, J., Feng, L., Han, Z., et al. 2015. Making place recommendations: An individual accessibility measure to urban opportunities in space and time. *Acta Geographica Sinica*, 70(6): 931–940 (in Chinese).

Hägerstrand, T., trans. Zhang, Y., and Chai, Y. 2010. What about people in regional sciences? *Urban Planning International*, 25(6), 10–17 (in Chinese).

Huang, X. 2009. A study on temporal-spatial behavior pattern of tourists based on time-geography science. *Tourism Tribune*, 24(6), 82–87 (in Chinese).

Huang, X. 2010. Time-geography and tourism planning. *Urban Planning International*, 25(6), 40–44 (in Chinese).

Huang, X. 2014. Study of conceptual framework of tourism temporal planning. *Tourism Tribune*, 29(11), 73–79 (in Chinese).

Huang, X. 2015. A study of tourists' emotional experience process based on space-time path: A case study of Ocean Park in Hong Kong. *Tourism Tribune*, 30(6), 39–45 (in Chinese).

Huang, X., and Ma, X. 2011. Study on tourists' rhythm of activities based on GPS data. *Tourism Tribune*, 26(12), 26–29 (in Chinese).

Huang, X., and Wu, B. 2012. Intra-attraction tourist spatial-temporal behaviour patterns. *Tourism Geographies*, 14(4), 625–645.

Huang, X., Zhu, S., and Zhao, Y. 2016. Product follows behavior: A tourism time product planning approach. *Tourism Tribune*, 31(5), 36–44 (in Chinese).

Kwan, M.-P., trans. Shen, Y., Zhao, Y., and Chai. Y. 2010. GIS methods in time-geographic research: Geocomputation and geovisualization of human activity patterns. *Urban Planning International*, 25(6), 18–26 (in Chinese).

Kwan, M.-P., Chai, Y., and Ta, N. 2014. Reflections on the similarities and differences between Chinese and US cities. *Asian Geographer*, 31(2), 167–174.

Lan, Z., and Feng, J. 2010. The time allocation and spatio-temporal structure of the activities of migrants in "village in city": Surveys in five "villages in city" in Beijing. *Geographical Research*, 29(6), 1092–1104 (in Chinese).

Lan, Z., and Feng, J. 2012. The spatio-temporal structure of migrants' daily activities of village in city: Case of typical villages in city of Beijing, China. *Scientia Geographica Sinica*, 32(4), 409–417 (in Chinese).

Liu, T. and Chai, Y. 2015. Daily life circle reconstruction: A scheme for sustainable development in urban China. *Habitat International*, 50, 250–260.

Liu, Y. 2005. *The Social Space of the Low Income in Urban China in Transition*. Beijing: Science Press (in Chinese).

Liu, Y., He, S., and Li, Z. 2005. Analysis of spatio-temporal structure of daily activities of urban poverty group in Nanjing. *Chinese Journal of Population Science*, S1, 85–93 (in Chinese).

Liu, Y., Liu, X., Gao, S., Gong, L., Kang, C., Zhi, Y., and Shi, L. 2015. Social sensing: A new approach to understanding our socioeconomic environments. *Annals of the Association of American Geographers*, 105(3), 512–530.

Liu, Z., and Chai, Y. 2001. A time-geographical framework in enterprise study: A re-explanation of the Taylor model. *Areal Research and Development*, 20(3), 6–9 (in Chinese).

Liu, Z., and Chai, Y. 2013. Danwei, family ties, and residential mobility of urban elderly in Beijing. In Besharov, D., and Baehler, K. (Eds.). *Chinese Social Policy in a Time of Transition* (pp. 196–222). Oxford: Oxford University Press.

Liu, Z., Zhang, Y., and Chai, Y. 2009. Home–work separation in the context of institutional and spatial transformation in urban China: Evidence from Beijing household survey data. *Urban Development Studies*, 9, 23 (in Chinese).

Liu, Y., Tan, Y., and Zhou, W. 2010. Japanese expatriates in Guangzhou City: The activity and living space. *Acta Geographica Sinica*, 65(10), 1173–1186 (in Chinese).

Ma, J., Chai, Y., and Liu, Z. 2011. The mechanism of CO_2 emissions from urban transport based on individuals' travel behavior in Beijing. *Acta Geographica Sinica*, 66(8), 1023–1032 (in Chinese).

Shaw, S.-L., trans. Yu, H., and Chen. J. 2010. A GIS-based time-geographic approach of studying individual activities and interactions in a hybrid physical virtual space. *Urban Planning International*, 25(6), 27–35, 44 (in Chinese).

Shen, Y., and Chai, Y. (forthcoming). Socio-spatial differentiation based on daily activity space. *Progress in Geography* (in Chinese).

Shen, Y., Chai, Y., and Kwan, M.-P. 2015. Space–time fixity and flexibility of daily activities and the built environment: A case study of different types of communities in Beijing suburbs. *Journal of Transport Geography*, 47, 90–99.

Sheng, X., Stockman, N., and Bonney, N. 1992. The dual burden: East and West (women's working lives in China, Japan and Great Britain). *International Sociology*, 7(2), 209–223.

Stockman, N. 1994. Gender inequality and social structure in urban China. *Sociology*, 28(3), 759–777.

Sun, D., and Chai, Y. 2017. Study on the urban community life sphere system and the optimization of public service facilities: A case study of Qinghe area in Beijing. *Urban Development Studies*, 24(9), 7–14 (in Chinese).

Sun, D., Chai, Y., and Zhang, Y. 2016. The definition and measurement of community life circle: A case study of Qinghe area in Beijing. *Urban Development Studies*, 23(9), 1–9 (in Chinese).

Ta, N., and Chai, Y. 2010. Time-geography and its enlightenments to human-oriented community planning. *Urban Planning International*, 25(6), 36–39 (in Chinese).

Ta, N., and Chai, Y. 2017. Spatial dilemma of suburban low-income residents: An analysis of behavior space among different income groups. *Acta Geographica Sinica*, 72(10), 1776–1786 (in Chinese).

Ta, N., Kwan, M.-P., and Chai, Y. 2016a. Urban form, car ownership and activity space in inner suburbs: A comparison between Beijing (China) and Chicago (United States). *Urban Studies*, 53(9), 1784–1802.

Ta, N., Kwan, M.-P., Chai, Y., and Liu, Z. 2016b. Gendered space-time constraints, activity participation and household structure: A case study using a GPS-based activity survey in suburban Beijing, China. *Tijdschrift voor Economische en Sociale Geografie*, 107(5), 505–521.

Tan, Y. 2017. *Research on Urban Socio-spatial Segregation and Ethnic Difference Based on Space-Time Behavior: A Case Study of Xining, China*. Unpublished doctoral dissertation, Peking University (in Chinese).

Tan, Y., Chai, Y., and Kwan, M.-P. 2017a. Examining the impacts of ethnicity on space-time behavior: Evidence from the city of Xining, China. *Cities*, 64, 26–36.

Tan, Y., Chai, Y., and Wang, X. 2017b. Space-time behavior of the Hui residents from the time-geography perspective in Xining city. *Areal Research & Development*, 36(5), 164–168 (in Chinese).

Tan, Y., Chai Y., and Kwan, M.-P. 2017c. Spatial variations of the geographic contextual effects on space-time behavior analysis: An empirical study in Xining, China. *Urban Development Studies*, 24(3), 22–30 (in Chinese).

Tan, Y., Chai, Y., and Kwan, M.-P. 2017d. The impact of the uncertain geographic context on the space-time behavior analysis: A case study of Xining, China. *Acta Geographica Sinica*, 72(4), 657–670 (in Chinese).

Wang, D., and Chai, Y. 2009. The jobs–housing relationship and commuting in Beijing, China: The legacy of *danwei*. *Journal of Transport Geography*, 17(1), 30–38.

Wang, D., and Ma, L. 2009. Simulation analysis of spatiotemporal distribution of visitors in Shanghai World Expo 2010. *Urban Planning Forum*, (5), 64–70 (in Chinese).

Wang, D., Zhu, W., Huang, W., et al. 2009. In-site visitor flow analysis and planning adjustment of the Shanghai Expo. *City Planning Review*, 33(8), 26–32 (in Chinese).

Wang, D., Chai, Y., and Li, F. 2011. Built environment diversities and activity–travel variations in Beijing, China. *Journal of Transport Geography*, 19(6), 1173–1186.

Wang, D., Li, F., and Chai, Y. 2012. Activity spaces and socio-spatial segregation in Beijing. *Urban Geography*, 33(2), 256–277.

Wang, D., Wang, C., Zhu, W., et al. 2015. Large-scale exposition planning and management optimization based on visitors' behavior simulation: A case study of Qingdao International Horticultural Exposition 2014. *City Planning Review*, 39(2), 65–70 (in Chinese).

Wang, Y. P., and Murie, A. 1999. Commercial housing development in urban China. *Urban Studies*, 36(9), 1475–1494.

Yin, Z. 2010. Directed movements in probabilistic time-geography. *International Journal of Geographical Information Science*, 24(9), 1349–1365.

Yin, Z., Sun, H., Chen, X., et al. 2015. Modeling uncertainty of directed movement via Markov chains. *Acta Geodaetica et Cartographica Sinica*, 44(10), 1160–1166, 1176.

Yue, Y., Lan, T., Yeh, A. G. O., and Li, Q. 2014. Zooming into individuals to understand the collective: A review of trajectory-based travel behaviour studies. *Travel Behaviour and Society*, 1(2), 69–78.

Zhang, C., Zheng, T., Bin, L., et al. 2009. Study on urban taxi drivers' out-dining characters and planning implications in Beijing. *Progress in Geography*, 28(3), 384–390 (in Chinese).

Zhang, W., and Chai, Y. 2008. Theories and confirmed model of urban residents' travel demand: Considering intra-household interaction. *Acta Geographica Sinica*, 63(12), 1246–1256 (in Chinese).

Zhang, X., Chai, Y., Chen, Z., and Tan, Y. 2016. Analysis of spatial and temporal patterns of daily activities of suburban residents based on GPS data: A case study of the Shangdi-Qinghe area of Beijing. *International Review for Spatial Planning and Sustainable Development*, 4(1), 4–16.

Zhang, Y. 2015. *Urban Spatial Behavior and Differentiation: Empirical Studies of Beijing*. Beijing: Academy Press (in Chinese).

Zhang, Y., and Chai, Y. 2011. The spatio-temporal activity pattern of the middle and the low-income residents in Beijing, China. *Scientia Geographica Sinica*, 31(9), 1056–1064 (in Chinese).

Zhang, Y., and Chai, Y. 2016. "New" time-geography: A review of recent progresses of time-geographical research from Kajsa Ellegård in Sweden. *Human Geography*, 2016(5), 19–24 (in Chinese).

Zhang, Y., Chai, Y., and Guo, W. 2014. Community differentiation of residents' daily activity spaces in Beijing city. *Areal Research and Development*, 33(5), 65–71 (in Chinese).

Zhao, Y., and Chai, Y. 2013. Residents' activity-travel behavior variation by communities in Beijing, China. *Chinese Geographical Science*, 23(4), 492–505.

Zhao, Y., Chai, Y. and Kwan, M.-P. 2015. Comparison of urban residents' travel behavior in China and the US: A case study between Beijing and Chicago. *Geographical Research*, 33(12), 2275–2285 (in Chinese).

Zhao, Y., Chai, Y., and Gui, J. 2016. Prospects for urban leisure studies in China: A perspective of space-time behavior. *Tourism Tribune*, 31(9), 30–40 (in Chinese).

Zhao, Y., Dijst, M., and Chai, Y. 2016. Between haven and heaven in cities: A comparison between Beijing (China) and Utrecht (the Netherlands). *Urban Studies*, 53(12), 2469–2487.

Zhao, Y., Wang, L., Huang, X., et al. 2017. The impact of audience participation at a theme park show on tourists' activity space: A space-time accessibility perspective. *Tourism Tribune*, 32(12), 49–57 (in Chinese).

Zheng, K., Jing, K. L., Jia, L. J., Hu, J., and Wang, Y. T. 2009. A study on temporal and spatial characteristics of the shopping activities of minorities in city: A case study of the Uygur residents of Urumqi. *Yunnan Geographic Environment Research*, 21(3), 16–21 (in Chinese).

Zheng, K., Cui, N., Li, Y., and Wu, T. 2011. A comparative study on hierarchy of shopping trip space: A case study of the Han and Uygur residents of Urumqi. *Yunnan Geographic Environment Research*, 23(4), 25–30 (in Chinese).

Zhou, S., and Deng, L. 2010. Spatio-temporal pattern of residents' daily activities based on T-GIS: A case study in Guangzhou, China. *Acta Geographica Sinica*, 65(12), 1454–1463 (in Chinese).

Zhou, S., Yang, L., and Deng, L. 2010. The spatial-temporal pattern of people's daily activities and transportation demand analysis: A case study of Guangzhou, China. *International Conference on Management and Service Science, IEEE*, 1–4.

Zhou, S., Deng, L., Kwan, M.-P., and Yan, R. 2015. Social and spatial differentiation of high and low income groups' out-of-home activities in Guangzhou, China. *Cities*, 45, 81–90.

Zhu, W., and Wang, D. 2008. Space choice behavior and multi-stop tracks of consumers in east Nanjing Road. *City Planning Review*, 31(3), 33–40 (in Chinese).

4 Green, healthy time-geography

Using time-geographic concepts for sustainable mobility research

Harvey J. Miller, Ying Song and Calvin P. Tribby

Introduction

Mobility has benefits: trading time for space in movement allows individuals to access activities and resources that are sparsely distributed across space for limited durations and times. However, this otherwise beneficial trade-off can have costs to society and the environment.

The automobile-dominated mobility systems in the United States and other parts of the world have created levels of human mobility and activity that are unprecedented in human history (Schafer and Victor 2000). However, these systems also have huge costs including bad health outcomes due to poor air quality and physical inactivity, climate change due to greenhouse gas emissions, depletion of non-renewable resources such as petroleum, high levels of injuries and death due to road trauma, wasted time and energy due to congestion, and social inequities due to the high costs and capabilities required to own and operate a personal motor vehicle (Black 2010; Carlsson-Kanyama and Linden 1999; Chapman 2007; Glover 2017; Rabl et al. 2005; Steg and Gifford 2005). Because of their high social and environmental costs, automobile-dominated mobility systems are not sustainable: they cannot last over the long run, especially as world population continues to grow and urbanize (Sperling and Gordon 2009). Transitioning to more sustainable mobility systems requires a reorientation of our approaches to transportation systems planning: away from facilitating mobility and towards managing mobility, including shifting people to substitute other forms of interaction for travel, taking shorter trips and using greener and healthier travel modes such as walking, biking and public transit (Banister 2008).

Time-geography offers a conceptual framework and toolkit that can help shift mobility systems planning from its conventional roles of predicting flows and providing infrastructure to shaping mobility demands towards more sustainable outcomes. Time-geography offers an individual-level perspective on mobility and activities, providing scientists and planners with alternatives to conventional measures of transportation systems performance such as traffic counts and flows. This chapter discusses time-geography and its role in facilitating sustainable mobility. In particular, this chapter highlights two research projects that use time-geographic techniques and perspectives to answer research questions surrounding more sustainable forms of travel. The first project – Green Accessibility: Measuring the Environmental Impacts of Network-Time Prisms for Sustainable Transportation Planning – develops methods for estimating the expected energy consumption and emissions associated with accessibility as measured using space-time prisms and network time prisms. It also contributes to the growing analytical foundation for time-geography by characterizing properties of the prism interior. The second project – the Moving Across Places Study – uses time-geographic principles to understand the relationships between public transit, the built environment and walkability.

The next section of this chapter briefly discusses the concept of sustainable mobility, and contrasts conventional transportation planning with sustainable mobility planning. The following section discusses the Green Accessibility project addressing the expected emissions associated with space-time and network-time prisms. The subsequent section discusses the MAPS project addressing active transportation (walking, biking and public transit). The final section concludes the chapter with some summary comments and research frontiers.

Sustainable mobility

What is sustainable mobility?

Sustainable mobility is a complex concept. Most people can agree that mobility offers benefits to individuals, societies and economies but can recognize that it also has negative impacts on people, communities and the environment. People can also agree that we should maximize the benefits but minimize the social and environmental costs of mobility. But what do we mean by "sustainable" mobility?

The most common definition of sustainability comes from the 1987 publication of the report "Our Common Future" by the World Commission on Environment and Development (WCED), also known as the Brundtland Report after its Chair, Gro Harlem Brundtland, former prime minister of Norway. This defines sustainable development as development that meets the needs of the present without compromising the ability of future generations to meet their own needs (WCED 1987). This identifies sustainability as a norm encompassing both intra- and intergenerational justice (Derissen et al. 2011). The Brundtland Report also argues that sustainability spans environmental, economic and social dimensions. We can view sustainable mobility in a similar manner: the ability to meet society's need for mobility and accessibility without sacrificing other essential human or ecological values, today or in the future. Sustainable mobility also spans environmental, economic and social dimensions: mobility should provide benefits of movement, access and interaction while minimizing environmental harms, being financially viable and socially equitable (Lagan and McKenzie 2004). We can argue about needs versus wants, and how many future generations is meaningful as a planning horizon, but the kernel of this definition is unequivocal: we should take an ecological approach to mobility planning, respecting limits, using a long-term perspective, and integrating the social and natural sciences in our attempts to understand and plan for more sustainable mobility. These are also key messages in time-geography (Ellegård and Svedin 2012).

Planning for sustainable mobility

How do we resolve or mitigate the unsustainable outcomes from mobility while still maintaining the benefits as much as possible? Part of the answer is shifting from the conventional "predict and provide" paradigm in transportation planning to one that focuses on managing rather than facilitating travel demand (Banister and Button 1992; Owens 1995). Banister (2008) contrasts the conventional approach to mobility planning with emerging approaches to sustainable mobility planning (see Table 4.1). Several features of sustainable mobility planning harmonize with the time-geographic perspective, in particular, an emphasis on accessibility rather than mobility, a focus on people rather than traffic, and local-scale rather than large-scale interventions. Time-geography can also provide insights with respect to the trade-offs of making time reasonable and reliable rather than minimal, and slowing movement down rather than speeding traffic up. Since time-geography focuses on people and activities rather than disembodied trips, it is also a powerful framework for understanding integrating modes such as walking and biking, as well as integrating people, activities and transportation.

More generally, time-geography and the activity-based approach to analyzing mobility demands are better suited for sustainable mobility planning than the classic trip-based approach to travel demand analysis. Since time-geography

Table 4.1 Contrasting approaches to mobility planning (based on Banister 2008; Marshall 2001)

Conventional planning	Sustainable planning
Physical	Social
Mobility	Accessibility
Traffic focus	People focus
Motorized modes	All modes, but emphasize walking and bicycling
Technocratic	Community-based
Modeling for prediction	Modeling for scenario development
Economic	Economic, environmental and social
Large scale	Local scale
Street as road	Street as space
Speed traffic up	Slow movement down
Time minimized	Time reasonable and reliable
Segregate activities, people, transportation	Integrate activities, people, transportation
Travel as derived demand	Travel as a valued activity as well as a derived demand

addresses the fundamental conditions for travel and activity participation, it is more effective for understanding travel demand management as well as non-transportation solutions to accessibility problems. Time-geography and activity-based approaches also harmonize better with emerging methods for estimating environmental impacts of transportation and understanding the relationships between transportation and public health. Finally, time-geography and activity-based approaches are better at accounting for diversity in the population; e.g., groups can be distinguished based on their activity patterns and constraints rather than their home locations. Therefore, they are better suited for examining social equity issues in sustainable transportation (Ellegård and Svedin 2012; Miller 2017).

Estimating the environmental costs of accessibility

Mobility researchers and planners use the space-time prism (STP) and the network-time prism (NTP) to delineate individuals' accessibility considering scheduling constraints and the ability of the given environment to afford the trading of time for moving in space. Prisms with larger volumes indicate higher potential mobility and therefore greater accessibility; hence, bigger prisms are usually considered better (Cetin and Sevik 2016; Farber et al. 2013; Lenntorp 1976; Widener et al. 2015). However, greater accessibility, if realized via automobile travel or other motorized means, can also bring greater environmental costs such as non-renewable resource consumption and greenhouse gas (GHG) emissions.

The Green Accessibility project develops a methodology to estimate the expected environmental costs associated with a prism through three consecutive steps: (1) simulating prism *visit probabilities* within the prism, which describes the probability for the vehicle to present at a particular location among all accessible locations within a prism; (2) deriving prism *speed profiles*, which describes the possible speeds when the vehicle passes those accessible locations; and (3) translate expected speeds into mobility-related environment costs such as energy consumption and CO_2 emissions using an *environmental cost function*. The project also develops procedures to calibrate the model using empirical movement data, and validates the model using secondary and primary data collected in New York City, NY and Phoenix, AZ in the United States. The rest of this section discusses the methods for each step, and provides example calibration and simulation results. In addition to applications in sustainable transportation analysis and planning, this framework and these techniques also add to the foundation for analytical time-geography by improving our understanding of the properties of the prism interior.

Visit probabilities within STP and NTP

In general, the prism is treated as binary: all locations within a prism are equally accessible while all locations outside the prism are equally inaccessible. However, the prism interior is not homogeneous: locations near the central axis connecting the prism anchors are more likely to be visited than those toward the prism boundary since there are more possible paths near the central axis than the boundary (Downs and Horner 2012; Lenntorp 1976; Winter and Yin 2010, 2011; Xie and Yan 2008). We model the visit probability distribution within the STP and NTP as stochastic processes constrained by the prism. For the STP, we adopt *Brownian bridge* (BB) techniques (Song and Miller 2014). For the NTP, we apply *Markovian techniques* (Song et al. 2016).

Modeling STP visit probabilities using Brownian bridges

To delineate the accessible locations an individual can present, we adopt the analytical definition of STP by Miller (2005). Given a pair of prism anchors, a maximum speed and a time budget, the spatial extent of the prism at a moment of time during the prism's existence is the intersection of at most two of three convex spatial sets: (i) a *future disc* encompassing the locations that can be reached from the first anchor given the elapsed time so far; (ii) a *past disc* that encompasses locations from that can reach the second anchor during the remaining time left in the budget; and (iii) a *potential path area* (PPA) that further constrains accessible locations given. For locations within the spatial extent of the prism at a moment in time, the times available for lingering before having to travel to the second anchor are unequal, implying unequal visit probabilities that are conditioned by available time. Also, the visit probabilities of all locations within the spatial extent must sum to unity.

To model the visit probabilities within STP, Winter and Yin (2010, 2011) use random walk theory. However, the resulting visit probability distribution and prism footprint has geometric artifacts not consistent with theory due to its discrete representations of space and time (Song and Miller 2014). We can improve the visit probability estimation by modeling space and time as continuous. We model movement within an STP as a *Brownian bridge* (BB): a continuous-time stochastic process that varies over time between two known values; in our case, movement between the prism anchors. For two-dimensional BBs, the probability distribution at any time follows a bivariate normal distribution with its expected/mean locations moving along the axis connecting two prism anchors at a designated average speed. We restrict the BB within the prism by applying a truncated normal distribution. We modify the standard deviation to ensure that the expected/mean location of the truncated distribution remains as the same as the one of the original distribution. We also include a mobility dispersion parameter to reflect how much an individual's movement is affected by the prism constraints (Song and Miller 2014). Figure 4.1 shows simulated visit probabilities at selected time step $t_k \in [0,100]$ for an STP. The light elliptical gray area indicates the prism PPA and the darker gray color toward the prism center indicates higher visit probabilities. Note the accessible locations have a changing spatial extent over time, with high visit probabilities located near the prism center.

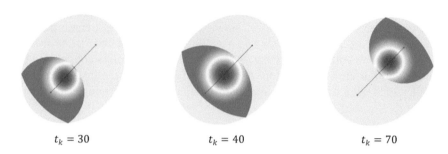

$t_k = 30$ $\qquad\qquad$ $t_k = 40$ $\qquad\qquad$ $t_k = 70$

Figure 4.1 Truncated Brownian bridges simulation of visit probabilities over time within a space-time prism

(Song and Miller 2014)

Modeling NTP visit probabilities using Brownian motion and semi-Markov processes

NTPs model accessibility for the case where individuals' movements are confined by spatial networks like streets, highways, biking lanes and pavements. We model the spatial network as a directed graph and we adopt the analytical definition of Kuijpers and Othman (2009); this defines the boundary of the NTP by the earliest possible arrival time from the first anchor and the latest possible departure time to the second anchor at locations within the network. To model visit probabilities within NTPs, we apply two Markovian techniques. For individual movements that are less restricted to the spatial networks (e.g. walking, biking), we model them as *Brownian motion on graphs* conditioned to geometries of the spatial networks and the achievable speed of that travel modes. For individual movements that are more restricted by spatial networks (e.g. driving), we model movement as *semi-Markov processes*, which are restricted to both geometries and properties of the spatial networks (e.g. speed limits, turning restrictions) (Song et al. 2016).

Figure 4.2 shows results from a Brownian motion simulation of visit probabilities for a 15-minute walking scenario within a portion of the Manhattan, New York City street network. Figure 4.2 also provides a planar STP PPA (gray circle) for reference. As expected, the accessible edges are changing over time and edges along shorter paths have higher visit probabilities.

For more constrained movement within networks, we model visit probabilities as *continuous-time semi-Markov processes*. This consists of two parts: (1) a transition probability that describes how likely the movement will be from a vertex to its connected vertices; and (2) the time required to complete this transition (Howard 1971). We modify the semi-Markov technique to consider whether a vertex is accessible, and how much time it required to pass through an edge. Therefore, we formulate a holding time density function to describe

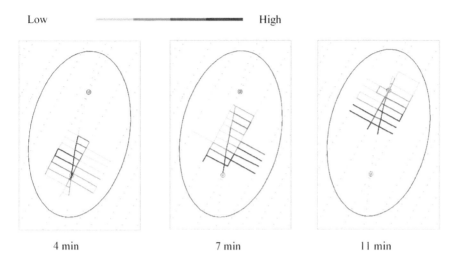

Figure 4.2 Brownian motion simulation of NTP visit probabilities over time

(Song et al. 2016)

the probability that the movement from a vertex to an adjacent vertex takes an extra amount of time beyond the minimum travel time. The holding time density function can be calibrated for a specific network using empirical trajectory data (Song et al. 2016).

Figure 4.3 shows simulated NTP visit probabilities for driving across lower Manhattan, NYC, USA in 25 minutes from the Holland Tunnel to the Manhattan Bridge. We calibrate this simulation using over 135,000 vehicle trajectories collected over two weeks by a commercial navigation company; this indicates that the holding times within this network follow an exponential distribution (see Song et al. 2016 for details). Figure 4.3 also illustrates that a substantial number of links theoretically within the NTP have a very low visit probability; these are the links with a less than 0.01% probability. This suggests that NTPs over-bound realistic accessibility, a new theoretical insight.

NTP speed profiles

When passing through an accessible location in a prism, at any moment a moving object may have a different range of possible speeds. Intuitively, we can see that locations near the center of the prism are likely to have a wider range of possible speeds that will still allow the individual to reach the second anchor by the required time. In contrast, an individual at the border of the prism must travel at the maximum speed to reach the second anchor by the required time. Therefore,

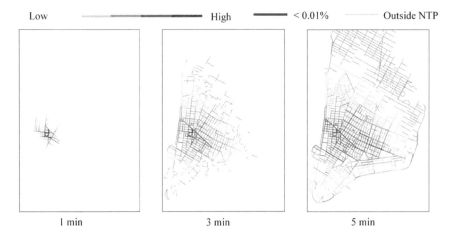

Figure 4.3 Semi-Markov simulation of NTP visit probabilities over time
(Song et al. 2016)

locations in STP and edges in NTP have different speed profiles or the range of possible speeds that meet the prism constraints.

We develop and compare three methods to derive speed profiles within a transportation network: (i) constant speed based on the expected arrival and departure times; (ii) constant speed based on the truncated arrival and departure times; and (iii) travel time along the edge following the holding time density function calibrated from empirical vehicle trajectories provided by a local ridesharing firm. Based on the empirical speed distributions derived from trajectory data collected in Phoenix, AZ, USA, we found that the holding time density method provides the best estimation of the speed profiles along accessible edges; Figure 4.4 shows the result (Song et al. 2017). We use the same holding time density function to calibrate NTP visit probabilities.

Using visit probabilities and speed profiles to estimate NTP environmental cost

There are various types of environmental and social costs associated with mobility that are a function of vehicle speed; these include fuel consumption, tailpipe emissions and road trauma. We focus on emissions and show how to use a NTP's visit probabilities and speed profiles to estimate the expected CO_2 emissions associated with that prism given a specified vehicle type.

We use MOVESLite, a simplified version of the US EPA's Motor Vehicle Emissions Simulator (MOVES) mode. The MOVESLite estimates the energy consumption and emissions by estimating vehicle specific power (VSP) (EPA 2012). This model can account for vehicle-specific characters including tire

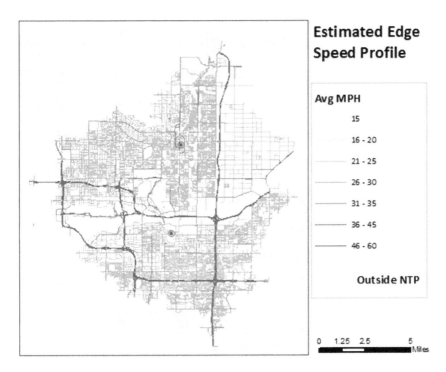

Figure 4.4 Estimated mean speed profiles for the Phoenix, AZ, USA road network
(Song et al. 2017)

rolling resistance, rotational resistance, aerodynamic drag coefficient and source mass factor mass, mobility-related parameters including the vehicle speeds and accelerations, and the road grade. MOVESLite model uses a VSP-to-Operation Mode conversion table to obtain average fuel consumption and emission rates stratified by the vehicle type, age and operation mode.

We define an NTP within the Phoenix road network, simulate an NTP's visit probabilities and speed profiles, and translate the simulation results into inputs for MOVESLite. We validate the emissions estimates from this modeling system using vehicles instrumented with GPS-enabled onboard diagnostic (OBD) devices that report engine operating parameters on a second-by-second basis; this allows empirical estimation of vehicle energy consumption and emissions. The fit between NTP emissions estimates using our method and the vehicle-derived emissions is good; see Song et al. (2017) for details on the integration method and validation procedures.

Figure 4.5 shows the simulated distribution of expected CO_2 emissions for the NTP within the Phoenix, AZ road network and how it changes over time. It indicates that the cost distributions are largely affected by the distributions of visit

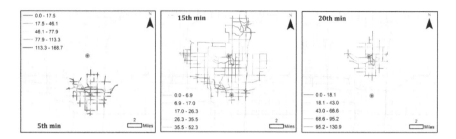

Figure 4.5 Distribution of expected CO_2 emissions (mg/s) at selected times within an NTP (Song et al. 2017)

probabilities, with higher costs along quicker routes. However, edges with higher speed limits also tend to bring more CO_2 emissions. In addition, larger accessible areas in Figure 4.5 always result in larger amounts of CO_2 emissions, which is consistent with our intuition that greater mobility may lead to higher potential costs to the environment.

Using time-geography to understand public and active transportation

Time-geography is a powerful perspective for understanding the factors that associate with use of public and active transportation, and the consequences of this behavior. From a sustainability perspective, a major policy question is how to get individuals out of private automobiles in favor of public transportation modes such as buses and rail and active transportation modes such as walking and biking (Glover 2017). From a public health perspective, greater use of active transportation modes is necessary to reduce the harm caused by lifestyle diseases associated with sedentary lifestyles and a lack of physical activity (PA) (Berrigan et al. 2006). Public and active transportation modes are complementary: public transit tends to pick up and drop off walkers and bikers rather than drivers (Besser and Dannenberg 2005; Brown and Werner 2007).

The section describes research that is a component of the Moving Across Places Study (MAPS): a quasi-experimental longitudinal study to assess PA and health outcomes before and after construction of a light rail transit (LRT) line and walk-ability enhancements in a neighborhood in Salt Lake City, Utah, USA. Participants (*n*=536) wore Global Positioning System (GPS) receivers and accelerometers for a minimum of ten hours per day for at least three days before the construction of the new LRT line (2012), called TRAX. Participants wore the equipment again approximately one year later, after the construction of the TRAX line (2013). Trained research assistants also administered attitudinal surveys, measured partici-pants' height and weight, and conducted high-resolution (block-level), multivariate

(160+ attributes) walkability field surveys in both time periods. Participants also delimited their neighborhood on study area maps and answered attitudinal questions about perceptions of walkability in these self-defined neighborhoods. Major results from MAPS suggest that the new TRAX line and walkability enhancements generated healthy changes in PA and body mass index (BMI) (see Brown et al. 2015; Brown et al. 2016).

We describe briefly three studies within the broader MAPS project that use time-geographic concepts and methods. These studies address the following questions. First, does new public transit generate new PA or just shift this time from other forms of PA, such as recreational walking? Second, how can we summarize walkability within behaviorally meaningful regions such as activity spaces? Third, which spatial scale is more meaningful for explaining home-based walking: self-defined neighborhoods or individual activity spaces?

Does public transit generate new physical activity time?

Evidence based on self-reports and cross-sectional analysis indicates that public transit users are more physically active than non-transit users (see, e.g., Freeland et al. 2013). However, it is unclear if public transit *causes* higher PA levels, for two reasons. One reason is confounding effects: high-quality public transit such as trams and light rail tend to be provided in neighborhoods that are dense and diverse. Neighborhood density and diversity may cause the higher PA levels, not public transit. A second reason for doubt is familiar to time-geographers: constraints imposed by individual time budgets. In many cities, trips by public transit take more time than the same trips by automobile. Since individuals have limited amounts of discretionary activity time, the additional time required for public transit trips may mean individuals have less time to spend in recreational PA such as walking the dog or neighborhood strolls. Because of these time constraints, PA time associated with public transit may be shifted from other PA time. While this is not harmful, it undermines arguments for public transit as a health benefit (Miller et al. 2015).

To test whether public transit generates new PA, we examine changes in PA time before and after the opening of the new TRAX line using the participants' GPS and accelerometer data. A semi-automated mode detection algorithm analyzes the fused GPS and accelerometer data and classifies observed trips with at least moderate levels of PA based on the travel mode; these are walk, bike, bus, automobile and LRT. Based on these results, we divide study participants into four groups. The *never* group is participants who did not use public transit during the week observation periods before and after the new TRAX line. The *continued* group used public transit at least once during both observation periods. The *former* group used public transit at least once before the new LRT service but did not use it after it opened, while *new* users did not use public transit before the new LRT but did use it at least once after. We also designate *transit-related PA time* as PA time associated with a trip that used public transit for at least one segment; *non-transit-related PA* is PA time associated with a trip that does not involve public transit. *Total PA time* is the sum of transit and non-transit PA time.

Table 4.2 summarizes the changes in PA time we should observe if public transit generates new PA. Our empirical results confirm these hypotheses: we find that participants who did not change their public transit behavior (Never, Continuing) did not have any significant changes to their total PA time, while the groups that did change their public transit behavior (Former, New) did have significant changes in their total PA time that is entirely due to changes in their transit-related PA time (see Miller et. al. 2015 for details). Therefore, we can conclude that the evidence supports the hypothesis that new public transit generated new physical activity time. Do these changes in PA time make a difference? Yes – we also find that the Former group experienced a significant increase in their body mass index (BMI) while the New group experienced a significant decrease in their BMI, using the Never as a control group (see Brown et al. 2015).

Figure 4.6 provides a density map of GPS points associated with moderate to vigorous PA for the New transit riders. Black squares correspond to new TRAX stations, black circles correspond to existing stations and black dots correspond to bus stops. From this map, we can see that three of the five new TRAX stops attracted new PA. The two new stops in the middle apparently did not generate new PA; these stops are situated near industrial uses and public land (a state fairground), illustrating that both new transit service and supportive land uses are required to generate new PA. Some of the existing TRAX stops and the bus routes feeding the new TRAX stops also generated new PA.

The results from this study provide supportive evidence for the health benefits of high-quality public transit. Time-geography provides insights that allows

Table 4.2 Research hypotheses about expected changes in physical activity time if public transit generates new physical activity (Miller et al. 2015)

	Change in behavior after new public transit service			
	Never used transit	*Continued using transit*	*Former transit users*	*New transit users*
(i) no confounding effects	No change in other (non-transit-related) PA time	No change in other PA time	Decrease in transit-related PA time	Increase in transit-related PA time
(ii) no time substitution			No increase in other PA time	No decrease in other PA time
Net change in total physical activity (PA) time	No change	Any change (due to possible change in transit-related PA only)	Decrease	Increase

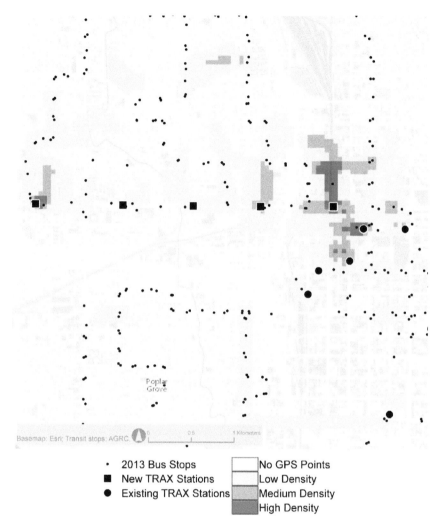

Figure 4.6 Density map of GPS points associated with new transit riders

(Miller et al. 2015)

untangling of the complex effects of transportation changes on individuals' allocations of time. Although promising, more longitudinal, quasi-experimental evidence is required.

Assessing walkability within activity spaces

There is growing evidence that supportive built environments have a positive influence on walking and bicycling (Brownson et al. 2009; Ding and Gebel 2012).

Walkability refers to the attractiveness and suitability of a built environment for walking (Ewing and Cervero 2010). Data for measuring walkability are frequently collected at two levels: (i) the census tract level; and (ii) the individual street block level. However, neither of these geographic regions are behaviorally meaningful: census tracts are too aggregate while individual blocks are too disaggregate. Required are methods for summarizing and assessing walkability within geographic regions that correspond to the built environment as experienced by individuals.

The related concepts of *potential path areas* and *activity spaces* provide meaningful geographic units for summarizing and analyzing walkability. The potential path area (PPA) is the limited portion of geographic space that is accessible to an individual during a given travel and activity episode. An activity space is the limited portion of the environment that a person experiences during regular travel and activities over some time period (daily, weekly, monthly, etc). Originating in time-geography and behavioral geography, respectively, they are closely related concepts with similar applications, with PPAs derived from space-time prisms and activity spaces derived from space-time paths (Patterson and Farber 2015).

We use activity spaces and PPAs as a meaningful basis for summarizing and analyzing walkability as likely experienced by individuals (Tribby et al. 2016). We identify walking trips from the participants' GPS trajectories to derive activity spaces. To derive a network-based PPA for each walking trip, we calculate the shortest path through the street network between the known trip endpoints and derive the PPA based on an assumed time budget. Using the street block walkability scores, we estimate summary statistics and network autocorrelation statistics for each activity space. We plot and map these activity space summary measures to compare the walkability of the built environment of these spaces within the study area.

We use three summary measures for assessing walkability within these activity spaces: (i) the average walkability score across street blocks (representing the general level of walkability in the activity space); (ii) the standard deviation (representing variation of walkability); and (iii) the network autocorrelation (representing the spatial coherence of the walkability pattern). Figure 4.7 illustrates a classification system for summarizing walkability based on these measures. The best situation is Quadrant I: this indicates high average walkability with low variability within the activity space. Conversely, Quadrant IV is the worst situation: low walkability with little variation within an activity space. The spatial autocorrelation of the walkability scores within an activity space can be positive or negative, reflecting the spatial coherence of the pattern. Positive spatial autocorrelation benefits walkability since spatial coherence helps individuals perceive and navigate walkable routes and regions.

Results indicate little correspondence between walkability summarized using census geography versus individual activity spaces. For example, Figure 4.8 compares the walkability summary scores measuring crime safety for census blocks and activity spaces; the small circles correspond to the centroid of the activity spaces computed in our study. The increased spatial detail of activity spaces reveals variations in the fine-grained summaries of the built environment that

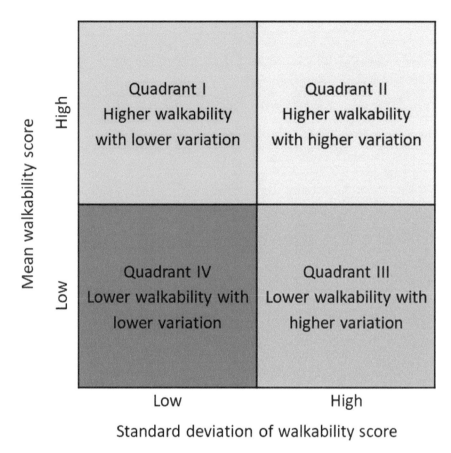

Figure 4.7 A classification system for summarizing walkability within activity spaces (Tribby et al. 2016)

coarse spatial units obscure. This suggests that aggregate measures of the built environment do not capture a pedestrian's exposure to the fine-grained diversity of built environment attributes. Time-geography provides a more meaningful basis for assessing individual experience with walkability in a built environment.

Neighborhoods, activity spaces and walking trips

Although activity spaces are meaningful for summarizing walkability, another meaningful geographic region for walking trips is the residential neighborhood surrounding a person's home. Researchers use both activity spaces and residential neighborhoods as regions to analyze walkability, with little agreement about which

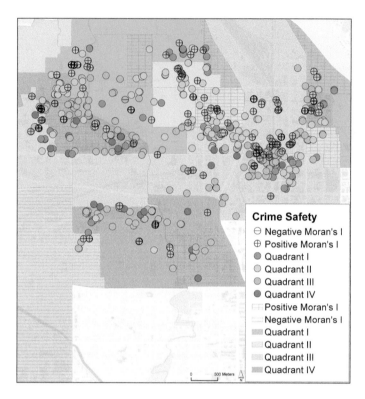

Figure 4.8 Comparing walkability summary measures for activity spaces versus
census blocks

(based on Tribby et al. 2016)

region is best for explaining these neighborhoods (Sharp et al. 2015). Using data
from MAPS, we can directly compare observed activity spaces with self-defined
neighborhoods for explaining home-based walking trips (Tribby et al. 2017).

The first step is examining the spatial relationships between self-defined
neighborhoods and walking activity spaces. We asked participants to free-hand
draw on a paper map the boundaries of what they consider to be their neighborhood; we digitized these maps and compiled them into a GIS database. We
derived activity spaces from participant GPS data by calculating a 200m buffer
around home-based walking trips and dissolving to create a polygon. Figure 4.9
shows an example of the spatial relationship between the two spaces for a selected
participant. For quantitative comparison, we compute geometric measures such
as the area, compactness and degree of geometric overlap between participants'
walking activity spaces and self-defined neighborhoods. Overall, participants'
neighborhood walks during a one-week observation period are only a small portion of their self-defined neighborhoods. We also examine changes in self-defined

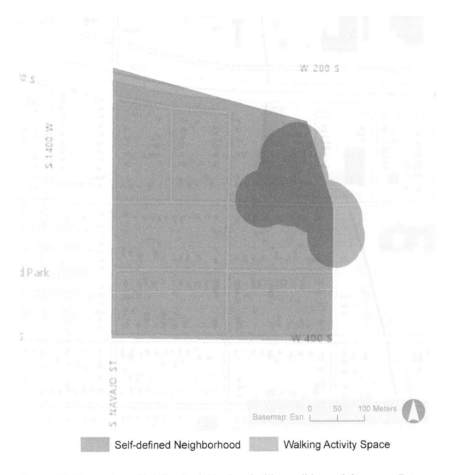

Figure 4.9 Comparing self-defined neighborhood with a walking activity space. Data are perturbed and generalized to ensure participant confidentiality.

neighborhoods and walking activity spaces over the two periods of observation approximately one year apart. We found that self-defined neighborhoods are stable over time, but walking activity spaces became more compact.

We also analyze the relationships between walking trips and the perceived and objective walkability qualities of neighborhoods, and changes in walking trips over the study horizon relative to changes in these qualities. We find that residents' exposure to objective measures of the built environment within activity spaces have a higher correlation with walking than the same measures for their self-defined neighborhoods. However, we also find that changes in both objective measures and perceived walkability in the self-defined neighborhoods are more important than these changes in walking activity spaces to explain changes in walking trips between 2012 and 2013. A possible explanation is that activity spaces encompass smaller areas than self-defined neighborhoods, and that these areas capture little variability in the built environment between the two years examined in this study. Overall, we find that both activity spaces and residential neighborhoods are meaningful behavioral regions for explaining walking trips, with the appropriate region depending on the questions asked and whether the analysis is cross-sectional or longitudinal.

Conclusion

Time-geography can support efforts to transform our currently unsustainable mobility systems to more socially and environmentally sustainable outcomes. In addition to other factors in the broader political economy, conventional transportation planning that focuses on predicting flows and providing infrastructure to accommodate those flows has led to contemporary mobility systems that harm environments, health and community. In contrast, sustainable mobility planning attempts to manage rather than facilitate travel demand, focusing on accessibility rather than mobility, people rather than traffic, and time as a resource rather than something that should be minimized at any cost. This chapter discussed research projects that use time-geographic concepts and methods to address question surrounding sustainable mobility.

The Green Accessibility project recognizes the environment costs associated with an individual's potential mobility represented by a space-time prism (STP) or network-time prism (NTP). The project develops a general framework for estimating the speed-related environmental and social costs of the STP and NTP. We plan to continue this research by considering another important factor in vehicle energy consumption and emissions, namely, acceleration. There are recent advances in incorporating acceleration limits into prisms (Kuijpers et al. 2017). We also plan to investigate other mobility-related costs such as the exposure to bad air quality and noise pollution, and traffic fatalities. In summary, the methods and approaches developed in this project can help transportation researchers and planners to evaluate the costs to facilitate individuals' mobility and support the benefit–cost analysis in sustainable transportation planning practices.

The Moving Across Places (MAPS) study applies time-geographic concepts to understand public transit, walking and walkability. MAPS examines the potential trade-offs associated with devoting scarce time to public transit rather than recreational physical activity and found the trade-off did not exist: people who switched to public transit found new time to devote to the physical activity required to access and egress transit. This project also advances analytical methods to summarize walkability within behaviorally meaningful geographic regions, namely, active spaces and potential path areas. We also compared activity spaces to self-described neighborhoods to assess the appropriate spatial scales to measure the associations of the built environment on walking trips and find that both scales are meaningful. MAPS exploited a planned change in the real world – the construction of a new light rail transit line – to design and implement a quasi-experimental design with objective measures that provides stronger support for causality than cross-sectional studies that rely on self-reports or attitudinal surveys. Greater capabilities for collecting relevant mobility data mean that these experimental opportunities are becoming more prevalent. Time-geography can help researchers leverage these opportunities by combining a fruitful theoretical framework with newly available data on human mobility.

Sustainable and healthy mobility requires two major changes in our transportation systems. Over the shorter run, it requires shifting individuals away from automobiles and towards public and active transportation. Over the longer run, it requires shifting our transportation modes away from fossil fuels and towards greener, less polluting forms of energy (Glover 2017). This chapter illustrates that analytical tools based on time-geography can inform policies and plans that address both of these transformations in our transportation systems.

Acknowledgements

Material in this chapter is based on research supported by the US National Science Foundation (Green accessibility: Measuring the environmental costs of space-time prisms in sustainable transportation planning; BCS-1224102) and the US National Cancer Institute, National Institutes of Health (Moving Across Places Study; 1R01CA157509-01).

References

Banister, D. (2008) The sustainable mobility paradigm. *Transport Policy*, 15, 73–80.

Banister, D. and Button, K. (1992) *Transport, the Environment and Sustainable Development*. London: Spon.

Berrigan, D., Troiano, R.P., McNeel, T., DiSogra, C. and Ballard-Barbash, R. (2006) Active transportation increases adherence to activity recommendations. *American Journal of Preventive Medicine*, 31, 210–216.

Besser, L.M. and Dannenberg, A.L. (2005) Walking to public transit: Steps to help meet physical activity recommendations. *American Journal of Preventive Medicine*, 29, 273–280.

Black, W.R. (2010) *Sustainable Transportation: Problems and Solutions*. New York: Guilford.

Brown, B.B. and Werner, C.M. (2007) A new rail stop: Tracking moderate physical activity bouts and ridership. *American Journal of Preventive Medicine*, 33, 306–309.

Brown, B.B., Werner, C.M., Tribby, C.P., Miller, H.J. and Smith, K.R. (2015) Transit use, physical activity, and body mass index changes: Objective measures associated with Complete Street light rail construction. *American Journal of Public Health*, 105, 1468–1474.

Brown, B. B., Smith, K.R., Tharp, D., Werner, C.M., Tribby, C.P., Miller, H.J. and Jensen, W. (2016) A Complete Street intervention for walking to transit, nontransit walking, and bicycling: A quasi-experimental demonstration of increased use. *Journal of Physical Activity and Health*, 13(11), 1210–1219.

Brownson, R.C., Hoehner, C.M., Day, K., Forsyth, A. and Sallis, J.F. (2009) Measuring the built environment for physical activity: State of the science. *American Journal of Preventive Medicine*, 36, S99–S123.e12.

Carlsson-Kanyama, A. and Linden, A.L. (1999) Travel patterns and environmental effects now and in the future: Implications of differences in energy consumption among socio-economic groups. *Ecological Economics*, 30, 405–417.

Cetin, M., and Sevik, H. (2016) Change of air quality in Kastamonu city in terms of particulate matter and CO_2 amount. *Oxidation Communications*, 39, 3394–3401.

Chapman, L. (2007) Transport and climate change: A review. *Journal of Transportation Geography*, 15, 354–367.

Derissen, S., Quaas, M.F. and Baumgärtner, S. (2011) The relationship between resilience and sustainability of ecological-economic systems. *Ecological Economics*, 70, 1121–1128.

Ding, D., and Gebel, K. (2012) Built environment, physical activity, and obesity: What have we learned from reviewing the literature? *Health and Place*, 18, 100–105.

Downs, J.A. and Horner, M.W. (2012) Probabilistic potential path trees for visualizing and analyzing vehicle tracking data. *Journal of Transport Geography*, 23, 72–80.

Ellegård, K. and Svedin, U. (2012) Torsten Hägerstrand's time-geography as the cradle of the activity approach in transport geography. *Journal of Transport Geography*, 23, 17–25.

EPA (United States Environmental Protection Agency) (2012) *Assessing the Emissions and Fuel Consumption Impacts of Intelligent Transportation Systems*. BiblioGov.

Ewing, R., and Cervero, R. (2010) Travel and the built environment. *Journal of the American Planning Association*, 76, 265–294.

Farber, S., Neutens, T., Miller, H.J. and Li, X. (2013) The social interaction potential of metropolitan regions: A time-geographic measurement approach using joint accessibility. *Annals of Association of American Geographers*, 103, 483–504.

Freeland, A.L., Banerjee, S.N., Dannenberg, A.L. and Wendel, A.M. (2013) Walking associated with public transit: Moving toward increased physical activity in the United States. *American Journal of Public Health*, 103, 536–542.

Glover, L. (2017) *Community-Owned Transport*. Abingdon: Routledge.

Howard, R.A. (1971) *Dynamic Probabilistic Systems*, vol. 2: *Semi-Markov and Decision Processes*. Chichester: John Wiley.

Kuijpers, B. and Othman, W. (2009) Modeling uncertainty of moving objects on road networks via space-time prisms. *International Journal of Geographical Information Science*, 23, 1095–1117.

Kuijpers, B., Miller, H.J. and Othman, W. (2017) Kinetic prisms: Incorporating acceleration limits into space-time prisms. *International Journal of Geographic Information Science*, 31, 2164–2194.

Lagan, C. and McKenzie, J. (2004) *Sustainable Cities, Sustainable Transportation*. EarthTrends. Available at: www.uky.edu/~tmute2/GEI-Web/GEI/GEI10/GEI 2010 assignments/GEI assignments/pop_fea_transport.pdf (accessed June 2018).

Lenntorp, B. (1976) Paths in space-time environments: a time geographic study of movement possibilities of individuals. *Environment and Planning*, 9, 961–972.

Marshall, S. (2001) The challenge of sustainable transport. In A. Layard et al. (eds.) *Planning for a Sustainable Future*. London: Spon, 131–148.

Miller H.J. (2005) A measurement theory for time-geography. *Geographical Analysis*, 37, 17–45.

Miller, H.J. (2017) Theories and models in urban transportation planning. In S. Hanson and G. Giuliano (eds.) *The Geography of Urban Transportation*, 4th edition. New York: Guilford, 113–138.

Miller, H.J., Tribby, C.P., Brown, B.B., Smith, K.R., Werner, C.M., Wolf, J., Wilson, L. and Oliveira, M.G.S. (2015) Public transit generates new physical activity: Evidence from individual GPS and accelerometer data before and after light rail construction in a neighborhood of Salt Lake City, Utah, USA, *Health and Place*, 36, 8–17.

Owens, S. (1995) From "predict and provide" to "predict and prevent"?: Pricing and planning in transport policy. *Transport Policy*, 2, 43–49.

Patterson, Z. and Farber, S. (2015) Potential path areas and activity spaces in application: A review. *Transport Reviews*, 35, 679–700.

Rabl, A., Spadaro, J.V., and van der Zwaan, B. (2005) Uncertainty of air pollution cost estimates: To what extent does it matter? *Environmental Science and Technology*. 39, 399–408.

Schafer, A. and Victor, D.G. (2000) The future mobility of the world population. *Transportation Research Part A: Policy and Practice*, 34, 171–205.

Sharp, G., Denney, J.T. and Kimbro, R.T. (2015) Multiple contexts of exposure: Activity spaces, residential neighborhoods, and self-rated health. *Social Science and Medicine*, 146, 204–213.

Song, Y. and Miller, H.J. (2014) Simulating visit probability distributions within planar space-time prisms. *International Journal of Geographic Information Science*, 28, 104–125.

Song, Y., Miller, H.J., Zhou, X., and Proffitt, D. (2016) Modeling visit probabilities within network-time prisms using Markov techniques. *Geographical Analysis*, 48, 18–42.

Song, Y., Miller, H.J., Stempihar, J. and Zhou, X. (2017) Green accessibility: Estimating the environmental costs of network-time prisms for sustainable transportation planning. *Journal of Transport Geography*, 64, 109–119.

Sperling, D. and Gordon, D. (2009) *Two Billion Cars: Driving towards Sustainability*. Oxford: Oxford University Press.

Steg, L. and Gifford, R. (2005) Sustainable transportation and quality of life. *Journal of Transport Geography*, 13, 59–69.

Tribby, C.P., Miller, H.J., Brown, B.B., Werner, C.M. and Smith, K.R. (2016) Assessing built environment walkability using activity-space summary measures. *Journal of Transport and Land Use*, 9, 187–207.

Tribby, C.P., Miller, H.J., Brown, B.B., Smith, K.R. and Werner, C.M. (2017) Geographic regions for assessing built environmental correlates with walking trips: A comparison using different metrics and model designs. *Health and Place*, 45,1–9.

WCED (1987) *Our Common Future*. World Commission on Environment and Development. Oxford: Oxford University Press.

Widener, M.J., Farber, S., Neutens, T. and Horner, M. (2015) Spatiotemporal accessibility to supermarkets using public transit: An interaction potential approach in Cincinnati, Ohio. *Journal of Transport Geography*, 42, 72–83.

Winter, S. and Yin, Z.C. (2010) The elements of probabilistic time-geography. *Geoinformatica*, 15(3), 417–434.

Winter, S. and Yin, Z.C. (2011) Directed movements in probabilistic time-geography. *International Journal of Geographical Information Science*, 24(9), 1349–1365.

Xie, Z. and Yan, J. (2008) Kernel density estimation of traffic accidents in a network space. *Computers, Environment and Urban Systems*, 32, 396–406.

5 Bringing the background to the fore

Time-geography and the study of mobile ICTs in everyday life

Eva Thulin and Bertil Vilhelmson

Introduction

The technological landscape of everyday life is changing rapidly. In the past ten years, always-online comprehensive mobile platforms (e.g. smartphones, tablets and laptops) have replaced internet use on place-bound computers as well as mobile social contacts limited to voice calls and texting via pre-smart mobile phones. Now the digital spheres of social media, internet-based information, consumption opportunities, entertainment, culture, etc., are constantly available online. This technological shift will likely transform the basic patterns of ICT use, among youth in particular, in terms of its "placing" in the socio-spatial and temporal contexts of everyday life.

Until the late 2000s, internet usage by young people was in practice a fixed activity, spatially as well as temporally, often tied to the home, to some extent to school, and conducted during leisure and alone-time (in an offline sense). Significant parts of after-school time, evenings and weekends were devoted to online activity on the computer, while pre-smart mobile communication was slipped between or within ongoing activities, filling up empty moments and passive time slots. Mobile use was often portrayed as allowing a kind of background awareness going on in parallel as the individual passed through time and space, engaged in various foreground activities. Smart and mobile platforms, however, are now radically changing the preconditions for online activity.

This chapter develops this theme by scrutinizing how current smart mobile ICTs are actually situated within the contexts of young people's daily lives. As time–space constraints on use are relaxed, online activity in general might well "turn mobile", becoming fragmented in time and more or less smoothly incorporated into a perpetually ongoing background. At the same time, emerging and enhanced online practices, for example, related to micro-coordination, semi-synchronous group discussions, and remaining virtually present and aware, necessitate a certain frequency and rhythm of background checking and shifts in order to be successful, putting new constraints on ICT use and time use in general. A conflicting ordering of offline and online, co-located and mediated activities and contexts thus emerges. An individual must increasingly juggle multiple online and offline contexts, being engaged in parallel projects associated with different constraints as regards attention, authority, social expectations and duration. A background flow of digital media might then more often shift into the foreground, taking time, effort and attention from other ongoing activities. This could result in conflicts, frictions and wrecked projects.

In this chapter we explore the incorporation of smart mobile technologies into young people's daily lives in the form of both background and foreground activities. We look at the strategies, priorities and tensions associated with shifts between background activities and a foreground setting in various temporal, social–spatial, and activity contexts. In doing so, we rely on an in-depth case study of young people's use of smart mobile platforms (i.e. predominantly smartphones and tablets), employing a time-geographic approach and research methodology. We enhance this understanding by referring to time-geographic concepts and methods in describing the role of online background activities in daily life. We also contribute by presenting tentative findings and suggesting issues that merit exploration in further research.

Approaching the active and intervening background

The role of ICTs as a kind of "background activity" is a recurring theme in debates about the internet, time use and everyday life, not least with the shift towards smart mobile platforms. This is linked to a wider discussion in conventional time-use research of the performance and significance of secondary activities in daily life, occurring in parallel with other activities. Not least, various types of "old" media usage have been shown to play significant roles as passive background activities, such as listening to music or the radio or watching TV while getting ready, cleaning, eating or hanging out (Robinson and Godbey 1997). A dominant hypothesis in the literature is that ICTs in general, and mobile ICTs in particular, increase the importance of simultaneity as a distinguishing feature of people's structuring and organizing of everyday life and time use (Urry 2000; Wellman 2001). Various concepts have been used to capture and enhance the understanding of this phenomenon, including notions of time-deepening (Robinson and Godbey 1997; Wajcman 2015), multitasking (Judd 2013; Kenyon and Lyons 2007), time-shifting (Wajcman 2015), and background awareness (Ito and Okabe 2005).

There are good reasons to believe that the shift toward smart mobile media and an increasingly wide range of always-accessible online opportunities will increase the significance of parallel activity. Yet less is known about how this enhanced online presence plays out and is practiced in the everyday life context, and with what implications. This lack of knowledge is partly explained by the fact that the phenomenon of secondary activity is regarded as elusive and difficult to observe and measure. Some insights can be drawn from previous studies of the "pre-smart" era of mobile ICT usage, focusing on certain segments of people's everyday use of time. Early studies of young people's use of mobile phones (via texting) for social contact describe such engagement as a form of background awareness, situated between contact and non-contact. Mobile phone usage is portrayed as continuously ongoing, but also transitory and almost invisible from a time-use perspective (Ito and Okabe 2005; Licoppe 2004; Ling 2004, 2013). Previous literature also highlights how mobile online activity goes on in parallel with passive activities and time uses. For example, conducting online activities (e.g. social contact, school work, entertainment and work) while traveling is found to be a common practice, often associated with perceptions of added value in terms of "task minutes" during a day (Aguliéra et al. 2012; Kenyon and Lyons 2007; Lyons and Urry 2005; Vilhelmson et al. 2011). Another example is when mobile digital media are used to fill slots of wasted time, such as waiting time, pauses, and "dead time", with online activity, contributing to the enhanced efficiency and densification of everyday time use (Bittman et al. 2009; Wajcman 2015). However, other studies examine parallel engagement with mobile digital media while performing activities that are not passive but require high attentiveness and concentration. Here we find a large body of research into young people's use of mobile ICTs and social media while studying (inside and outside classrooms) and into the habits and implications of multitasking. These studies find that heavy ICT users multitask to a significantly greater extent than do others (Janco and Cotton 2012; Judd 2013). Simultaneity in this context does not seem to represent added value or enhanced efficiency, but rather negatively affects the foreground activity. For example, school tasks take longer to complete and are of poorer quality when undertaken in a multitasking mode. There is also no evidence that young people ("the net generation") are better at multitasking than are other population segments. By focusing on processes of activity fragmentation, which is how online pursuits fragment ongoing purposeful sequences of action, some studies also note the practical difficulties of doing several things at the same time (Hubers et al. 2008, 2015). In essence, these insights suggest that the enhanced presence of an ongoing background of digital media is associated with highly variable levels of friction in everyday life, depending on the activity contexts, combinations and priorities.

In order to approach an understanding of how the practices and implications of simultaneity evolve in the context of current smart mobile technology, we advance three important notions concerning the enhanced role of time-geography, the flexibility and resistance of foreground activities, and the changing nature of online backgrounds.

First, we argue that the basic time-geographic model, describing the individual's daily life as an unbroken sequence of primary activities (offline as well as online), could be complemented by including a description of background time use comprising bundles of online activities performed in parallel with primary activities. These bundles intervene more or less, actively or passively, in ongoing foreground activity. This way of noting or considering the time used for background activity is in one sense a way of "double accounting" for everyday time use. The purpose of this method is not to violate the 24-hour time budget axiom, nor is it to dispute whether the individual has a limited capacity to actively engage in more than one activity at a time (Hägerstrand 1970; Schwanen and Kwan 2008). This approach is one way of capturing and visualizing how the presence of mobile media is changing the character of everyday time use, beyond the obvious fact that it is increasingly spent on online activities. The goal is to scrutinize combinations as well as potential conflicts and synergies between foreground and background activities.

Second, when developing a time-geographic understanding of parallel activities and the implications of the background use of mobile digital media, we believe it is crucial to pay attention to the various elasticities built into the unbroken sequence of activities. More specifically, previous studies indicate that the timing and frictions of parallel engagement in online and offline activities are highly related to the flexibility, resistance and interchangeability of the actual activities and projects of everyday life (Vilhelmson et al. 2017). For example, it is found that online background activity is sometimes slipped between foreground activities, time-shifted to natural pauses, conducted in parallel with passive foreground activity, or conducted concurrently with activities that are easy to interrupt, shorten and reorder. In this way the daily use of time becomes densified, yielding added value and increased efficiency. Previous insights also indicate, however, that engagement in online background activity is associated with intrusions and interruptions, causing distraction and unwanted congestion and stress in everyday life.

Third, we argue that the shift towards smart mobile media changes the nature of the background, implying a challenge and a need to complement the concept and notation of time-geography (Thulin et al. forthcoming). From a theoretical perspective, the constant presence of smartphones means that the background of online activity has not only been intensified, but is also becoming increasingly active and intervening in the foreground. This is especially true for online social projects that contain dense coupling constraints and that follow their own inherent rhythms of mediated interaction and co-attentiveness. As argued in the following section, a more active, semi-synchronous and intrusive background of online mobile media is not easily time-shifted and, in principle, intervenes in any type of activity, time and place. It potentially involves increasing levels of tension and ambivalence when contexts collide and projects are delayed or wrecked as the individual passes along the "now line" (Hägerstrand 1985).

A case study of ICT use in an era of smart mobile platforms

Time-geographic diaries capturing foreground and background use

Empirical research is scant concerning both how mediated activities actually influence everyday life, and what strategies people apply to manage ever-present online contact. This is partly because mobile ICT use is problematic to observe and register. Since smartphones are almost always on and are well integrated into people's everyday activities, their use is difficult to detect using regular methods for collecting time-use data and challenges the notation method of time-geography as well. The development of new and creative approaches to data gathering is therefore much needed. In this section, we make an effort to advance both time-geographic understandings and data collection methods. By conducting a case study of young people who are highly experienced smartphone users, we take a step towards improving the notation of mobile ICT use as both a foreground and a parallel activity.

Our data collection study extends the time-geographic diary method (e.g. Ellegård 2000; Palm and Ellegård 2011) to capture online time use – performed in the foreground as well as the background. The diaries are structured in terms of five core dimensions of daily activity: (i) *when*, i.e. at what time and for how long, activities were performed; (ii) the *content* of the *foreground activity* performed, offline as well as online; (iii) the *content of any parallel online activity* performed in the background; (iv) *where*, i.e. at what location activities were performed; and (v) with *whom*, i.e. the presence of co-located persons, offline. In the study, "parallel activities" refers to activity combinations performed during certain periods of time as defined by each respondent; the respondents describe these in connection with each foreground activity, whether or not it is combined with a background online activity. The duration of background activity is here defined by the time use of the associated primary activity.

We selected a group of 18 high school students living in the city of Gothenburg (Sweden), 10 women and 8 men, 17–18 years old, to keep a diary for three days (two weekdays and one weekend day). The diaries were distributed in the form of a physical notebook, but diaries could also be kept in an electronic version. Soon after the chosen days were recorded in the diaries, in-depth interviews with the participants were conducted by one of the authors. Data from the diaries were then used to describe the participants' time use, activity content and sociality at various locations. The interviews provided information about individual priorities, strategies and perceived frictions concerning foreground and background activities. The study was conducted in spring 2016. (For a more detailed description of method and data, see Thulin 2017.) It is important to note that the intention with the diaries was not to map in detail the individual acts of online background use of smartphones during the day, or the frequency and exact duration of each SMS, snap, or "like", for example. Rather, we wanted to target how ICTs influence and are integrated with the daily use of time and space by capturing how sequences of online background use interact with, interfere with or support any foreground activity. The diaries were coded for conventional group-level analysis and also treated at the individual level (e.g. analysed in terms of trajectories).

Group-level information on online time in the background and foreground

Initially, the diary-based information was used for a comprehensive description of the aggregate patterns, interactions and contexts of respondents' ICT use at the group level (see Table 5.1). The students spent almost 5 hours a day on online foreground activities, almost all using mobile platforms, i.e. mainly smartphones and to some extent tablets. Besides being an increasingly dominant primary activity of everyday life, online use in the background accompanied almost 7 hours of foreground activities per day.

Table 5.1 Online time use for various activities/purposes, both foreground and background (participants, $n = 18$)

	Foreground activity (n = 317)	*Background activity** (n = 533)
Total time online		
Time per person and day	4.7 hours	6.7 hours
Share of waking time	28.4%	40.8%
Type of online contact	*Minutes per person per day*	*Minutes per person per day*
Social contact		
Conversation (voice, Skype)	5	11
SMS, email	3	43
Social media	48	248
Information		
Info – personal interest	11	3
Info – news, traffic, weather	2	5
Info – school	11	10
School work	23	0
Shopping, booking tickets, *payment*	1	5
Entertainment		
Music	9	62
Film clips	8	4
Films, series	116	9
Games	32	4
Total	269	403

* Time spent on a background activity is defined and calculated by the duration of the associated, parallel foreground activity.

Of the online foreground activities, about 4.5 hours per day were spent on free-time activities, while only 30 minutes were spent on required activities, with school work being performed at school or some other place. Over 2 hours of use concerned entertainment in various forms, in particular watching TV series and films on mobile platforms, while time spent watching traditional TV has lost in importance. Half an hour was spent gaming. Online social contact constituted another substantial part of online time, almost 1 hour per day, as a foreground activity. This comprised interaction with friends and to some extent also with family. Online social time was completely dominated by social media, which provided visual dialogues (e.g. via Snapchat), group conversations (e.g. via Messenger), and social networking (e.g. via Instagram) (see Thulin 2017 for further analysis). Observably, very little total time was used on conventional SMS and voice calls, in line with other findings that these media have moved away from the center of young people's communication repertoire (Bertel and Ling 2016; Bertel and Stald 2013; Thulin 2017). Only about 10 minutes per day were devoted to information seeking in relation to personal interests, traffic and weather conditions, while no time was spent on news consumption.

The diaries indicate that time spent online, in social communication, on entertainment and being distracted above all, continues to be a highly prioritized foreground activity among young people. This clearly indicates that ICT use defined by purpose continues to be a meaningful, distinct and relevant foreground activity, despite concerns that mobile platforms and always-online access dissolve and blur the online/offline and primary/secondary distinctions.

However, blurring and distinction become more of an issue when we consider how mobile online media, besides expanding heavily in the foreground, constitute a highly active background of daily life. In the group of studied students, the time during which ongoing foreground activities were shared with online communication in parallel totaled almost 7 hours a day. Of course the intensity of interaction between foreground and background can vary greatly (as evidenced by the follow-up interviews), ranging from handling infrequent person-to-person messages to intensive exchanges and conversations with many people simultaneously online. Most importantly, the diaries showed that the online background was entirely dominated by social contacts – use of social media in particular – constituting 4 hours of background activity every day. The online background can therefore largely be described as bundles of the mediated co-presence of absent friends (Thulin et al. forthcoming). Furthermore, the impression of the background as active and intervening rather than passive and nonintrusive is reinforced. Yet certain background activities could still be deemed passive, listening to music being a typical example, constituting 1 hour of background activity per day.

The diaries also provide insight into the context of mobile ICT usage as regards space, sociality and associated activities, as regards where, with whom and in what combinations online activities were performed in the foreground and background. Importantly, we found that almost all primary foreground online activities took place in the home (see Figure 5.1). Obviously, ICT use continues to be a time-consuming, largely stationary and sedentary activity even when totally performed on mobile platforms. This observation opposes the hypothesis that mobile platforms lead to increasingly mobile use, and that free-time activities attached to the home are being replaced by spatially unconstrained and foot-loose out-of-home activities. As expected, background ICT use is more mobile in the sense of being performed at many different locations; however, the home is still the most common place even for background use, although much is also performed in school, when traveling, in public spaces, etc. As regards social context, online foreground time is largely a physically solitary activity. Almost 80 percent of all online activities in the foreground took place when the individual was alone, no other person being in physical proximity, compared with 50 percent of all offline activities. Also, online background activities were somewhat more social, 50 percent of them being performed with other people being co-present offline.

A core type of information from the diary concerns how online background activities were combined with the individual's sequence of foreground activities. This concerns what activity combinations were common, particularly how an insistent background interacted or interfered with certain foreground activities,

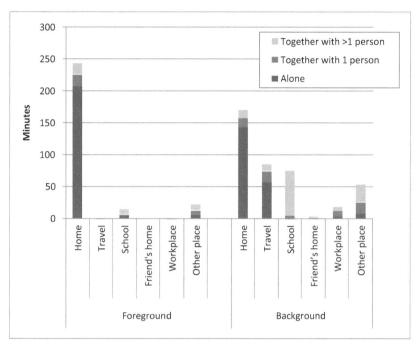

Figure 5.1 The location of online activity, in the foreground and background. Performed at various places, alone or together with other people (offline). Minutes per person per day; *n* = 18.

and, on the other hand, what activities were kept "free" from combination with or disturbance from background activities. At the individual level, we found a pattern indicating that activity combining was not continuous, but followed a particular rhythm – constrained or encouraged by ongoing foreground activities (as further discussed in the next section).

At the group level, nearly all everyday activities, starting from the time people got up in the morning and got ready for the day, ate, went to school, studied at school, took breaks, spent time with friends, etc., were more or less combined with a digital background of social contacts, social media in particular, but also texting and calling (see Figure 5.2). The intervening background was particularly dominant during the times of the day characterized as "elastic activities" – that is, activities less fixed in time, space and priority and perceived as likely to be cut short, interrupted, relocated or reordered. Physical movement, for example, by public transport, has been transformed into active online time (comprising parallel physical and virtual mobility) rather than passive offline waiting to arrive at one's destination. Pauses between lessons at school have become time for online socializing, filled with social media attention, rather than offline social interactions with classmates. Accordingly,

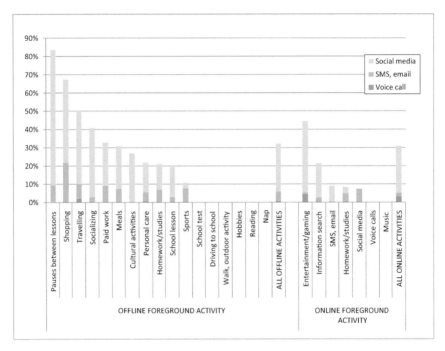

Figure 5.2 Online background activities associated with foreground activities. Ratio of online background time/foreground activity time.

time use and activity have become densified in line with early observations of time-shifting, time-deepening and filling empty moments. Yet it appears that even less-elastic activities, such as offline socializing with friends and family, school lessons and meals, have also been mixed with an active online social background. This implies that people have increasingly become involved in multiple contexts of presences associated with diverse expectations, splitting demands for attendance and attentiveness, and thus increasing the risks of friction, conflict and tension.

Overall, the aggregated diary data suggest that existing time–space patterns of ICT use as a time-consuming, solitary and stationary activity primarily linked to the home will not be drastically changed by the spread of smartphones and other mobile platforms. When ICT use is observed as a foreground activity, it is still fairly straightforward to capture using conventional methods, being linked to crucial questions about people's time priorities, sedentary behavior, loneliness, outdoor activity, news consumption, virtual mobility, travel, etc. The big challenge compared with the time before the advent of mobile devices is that of adding new layers of simultaneous online background that call for attention and are active, intervening and constraining. Indeed, this background is difficult to observe and register, prompting a need to elaborate on the time-geographic approach for description and understanding.

The rhythm of online social contact in young people's lives

The emergence of the semi-synchronous background of online social contact

Our knowledge of how intensified online sociality intervenes in and restructures everyday life is further enriched using a time-geographic conceptualization and individual-level approach. Of particular importance, then, is enhancing our understanding of the rhythms of online social contacts – or bundles of mediated co-presence – now underlying most young people's social lives and projects. Rhythm here denotes the frequency, intermittency, duration and, not least, intensity of social contact that binds and structures individual attention and time use over a day (Thulin et al. forthcoming). These rhythms are not freely situated in time–space but are associated with coupling constraints that bind and reorder everyday uses of time and place.

It is important to recall that the use of new mobile media represents a fundamental change in everyday online sociality. Smartphones have very quickly become integrated as a taken-for-granted and indispensable part of young people's everyday social activities and projects (Ling 2013; Thulin 2017). Their use adds new characteristics to the functions and meanings of mobile social contact initially established in the first wave of pre-smart mobile phone adoption (Kasesniemi and Rautiainen 2004; Licoppe 2004; Ling and Yttri 2002; Oksman and Turtiainen 2004; Thulin and Vilhelmson 2007, 2009). Now online social contact is increasingly visual and image based. Visual messaging, or fast dialog-like exchanging of images (of oneself, people one is with, places one is visiting and activities one is engaged in), is used to continuously share moments, situated experiences and emotions throughout the day (Bayer et al. 2016; Katz and Crocker 2015; Thulin 2017). Furthermore, mobile online contact is increasingly group oriented, taking place online in "mobile Messenger groups" of various sizes, reflecting the social circles of which one is part (e.g. closest friends, friends from one's old school, larger friendship networks and one's class) (Bertel and Ling 2016; Ling 2017; Thulin 2017). These groups are a main tool for planning and micro-coordinating joint activities with different time horizons and also serve as always-open meeting places where one can hang out, talk and socialize with friends. Moreover, mobile contact is no longer associated only with small circles of close friends and family members, but now often includes communication with networks of weaker social ties as well (Ling 2017; Ling and Lai 2016).

A tendency of particular importance from an everyday life perspective is for emerging "smart" mobile practices to be associated with changes in the rhythms and bonds of online social contacts, as regards frequency, interval, persistence and degree of synchrony. Whereas text-based person-to-person contact (i.e. SMS) is considered an asynchronous form of contact, liberating the user from the constraints of mutual co-attentiveness and simultaneous communication and easily time-shifted to moments of empty time (Kwan 2002; Ling 2004; Wajcman 2015; Yin et al. 2011), we are now witnessing the emergence of a near- or semi-synchronous discourse with profound effects on individual attentiveness and time use (Rettie 2009; Thulin 2017). This semi-synchrony is characterized by social contacts that are neither discretionary nor fully mutually attentive. It is coupled with increased social expectations and an inherent logic and rhythm of immediacy and responsiveness. Visual messaging is, for example, associated with a high expectation of very quick replies in a dialog-like manner. Increasingly real-time group conversations are characterized by extended and semi-concentrated sequences of co-attentiveness. This semi-synchrony is also experienced through episodes of waiting and inattention, sometimes creating frustrations as one is left hanging, and discussions become unfocused and unnecessarily prolonged. In essence, this semi-synchrony entails mutual expectations to pay conscious background attention to what is going on, said, shared and discussed. This return of synchronous communication also shows signs of increased intrusiveness and tendencies for the recoupling of contact; this stands in contrast to a dominant discourse concerning the capability of mobile ICT to relax time–space constraints (Schwanen et al. 2008; Yin et al. 2011).

Viewed from a time-geographic perspective, the present findings suggest that the rhythms of mediated social interaction are not only becoming more frequent and diverse but also are being recoupled and resynchronized (Thulin et al. forthcoming). These rhythms comprise bundles of mediated co-attentiveness (between two or more individuals) perceived to be just as real and "constraining" as co-located forms of social bundling and interaction. Importantly, the semi-synchronous flow of online mobile media is also likely to intervene in and reorder the unbroken sequence of everyday activities: The online background is becoming increasingly insistent and demanding of attention.

Strategies for juggling foreground and background activities

There is limited knowledge of how the intensified, more diverse and semi-synchronous background of online social contact is handled in practice. Starting from the time-geographic diaries, we obtained preliminary insight into how the daily rhythms of mediated co-presence are inserted into the sequence of daily activities and routines at different locations. Findings suggest the existence of highly diversified trajectories and patterns at the individual level, indicating different priorities as regards the domination versus subordination of online social contact when fitted into an everyday life context. From the diaries we derive three strategies for juggling background and foreground activities: the (i) *always ongoing*, (ii) *disciplined*, and (iii) *constantly renegotiated* bundling of activities and contacts with absent friends online.

Often we find that the young individual's engagements in online social contact are *always ongoing and detached* from the sequence of co-located contexts. This suggests that mediated co-presences among friends follow their own inherent logic and are arranged in accordance with separate and distinct online pockets of order (Thulin et al. forthcoming). In other words, absent friends are always allowed to "be there" in the online background, and the flow of social media in principle is allowed to intervene in any activity, time, place and social context.

This affects both elastic and less-elastic activities, time alone and with others, time online and offline, at home and elsewhere.

One typical example (illustrated by trajectory A in Figure 5.3) is person A who continuously communicated with absent friends through visual messaging (via Snapchat) and group conversations (via Facebook Messenger) in particular, and also by viewing and confirming the flow of images in his larger network of friends (via Instagram). His rhythm of online contact appears generally disconnected from – or not subordinate to – the activity sequences and routines of everyday life. He was co-present with friends online in the morning at home, when traveling to school, when in a lecture, during school breaks, and when spending time with his girlfriend in the evening. The only clear exception on this particular day (i.e. the activity category "free of digital background") was time spent on meals and self-care.

While this type of always-online bundling trajectory was common in our case study, we also found examples of individuals with the opposite pattern of fitting online social contacts into everyday routines. In such cases, the rhythm of mediated co-presence was clearly *linked to and disciplined by* the sequence of foreground activities, suggesting that it was subordinate, restricted and not allowed to flow freely. In particular, periods of passive time and time spent performing elastic activities (i.e. not as sensitive to interruption) were combined with mobile social media, for example, when traveling, on breaks from school, when alone at home, and at times of rest and entertainment. An example of this is the trajectory of person B, clearly showing a rhythm of mediated presence being "turned on and off" depending on the parallel foreground activity in focus. This person was virtually co-present with her friends, on Snapchat and Instagram in particular, during slots of waiting time, breaks, pauses and when watching online series by herself at home. Time for school work in particular,

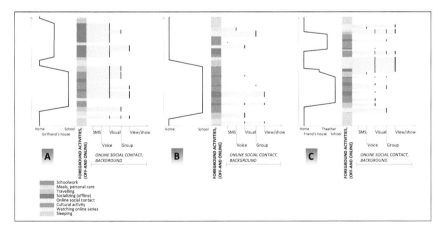

Figure 5.3 Typical strategies for managing background and foreground activities. A time-geographic representation of an always ongoing (A); a disciplined (B); and a constantly renegotiated (C) bundling of contacts with friends online.

both at home and when in school, was definitely free of digital background, as was social family time in the evening.

We also found individuals with less consistent or regular trajectories and patterns of foreground–background interaction. Their interaction could instead be seen as a continuously *ongoing negotiation and renegotiation* about when and where to be present with friends online, and about how to prioritize mediated versus co-located contexts. This is particularly apparent in the trajectory of person C, which represents a more fragmented and irregular pattern of mediated co-presence with friends. For example, parts of time for school work were kept free of digital background, while other parts were characterized by extensive parallel digital activity.

Negotiating the background

The activity and media-use trajectories reveal different versions of how mediated and co-located contexts in practice play out and are juggled in the everyday life context. Using in-depth interviews, the understanding of this interplay was enriched by exploring underlying negotiation strategies in greater depth. We found two particularly important strategies among those investigated. A first is to strive for *the smooth simultaneity of the digital and geographically co-located contexts* (i.e. foreground and background). The mediated presence of geographically absent friends is handled as effortlessly as possible, slipped almost unnoticed into and between ongoing activities. In this strategy, it is always acceptable to use one's mobile as long as one does so efficiently and smoothly. Importantly, the emerging smart practices of visual messaging and group communication both intensify and enable a "smarter" fitting of background contact into daily activities and routines. They sophisticate and "tame" the practices involved in staying socially connected and aware. Posting images and having visual conversations are more effortless, context revealing and quicker than writing and texting. Group communication is more seamless and eases the burden of having constant connection, as the responsibility for not breaking the connection is put on the group instead of the individual.

A second dominant strategy is essentially *to self-control the ongoing background* and from time to time "turn off", absenting oneself from online friends and dissolving the bundles of online co-attentiveness. In time-geography terms, this is a way to make the walls of the pocket of local order less porous (Lenntorp 2005). An important observation is that this was not done randomly or impulsively by the participants. Periods of absence and dropping out were always associated with negotiations within the social network. In practice, network members put considerable effort into informing one another in advance about not being present online or disappearing for a while, explaining that expected answers might be delayed, and by saying "good night" and "good morning". In the same vein, valid excuses were given for online absences, for example, that one had to study, it was family dinner time, a lesson was in progress or one was at the theatre and had to turn the mobile off.

Frictions in the juggling of online and co-located presences

Overall, we found that the studied young people were not victims of technology but highly active in developing various use strategies and in handling the flow of mobile social media on a daily basis. These strategies were very much negotiated in the social context of friendship networks, which could differ greatly. Even among this quite small group of studied individuals, it was possible to identify quite distinct "micro cultures" as regards mobile ICT use and personal availability, including different rhythms of contact and informal rules and expectations regarding presence and absence. However, while mediated co-presence was actively negotiated, it was also clear that the massive use of digital media – with its inherent expectations, rhythms and orders – has imposed new constraints on daily life. This implies that the juggling of mediated and co-located presences in the foreground and background periodically also engenders conflict, ambivalence and unease.

One common participant experience was that the mobile phone would quite suddenly shift into the foreground of attention and activity, even when the actual intention was to do something else. Furthermore, the participants could easily end up using the mobile device for considerably longer than initially intended. Even when they knew that the device should be turned off, doing so was often perceived as not that easy, sometimes being associated with stress and anxiety at not living up to expectations, saddening a friend, and missing out on things. There was also considerable tension and conflict in relation to other people's mobile media usage in co-located contexts. This was notably apparent in the classroom, where mobile phones were often perceived as very disruptive, taking time, energy and attention away from school work. Tensions also arose in other social situations, when the mediated presence of absent friends interfered with co-located quality time spent with physically present friends (e.g. when meeting at a cafe or having a dinner party). We therefore found that managing geographically co-located and mediated contexts respectively was far from frictionless and could occasionally be described as a clash between what is going on in foreground and background pockets of local order.

Conclusion

In summary, this chapter has discussed how the rapid spread of always-online smart mobile ICTs has transformed everyday life. We argue that an increasingly active online background of social contact intervenes in the ongoing sequences of the foreground activity of individuals. Sometimes background activities supersede the foreground, sometimes they are performed in parallel with it, and sometimes they are stopped altogether. The emergence of interfering backgrounds challenges the study of everyday life, as it is difficult to register their varying occurrence and character and challenging to understand their implications for individual action and attention. We tentatively explored the sophisticated use of a time-geographic perspective and diary notation method to examine both foreground and mediated background activities. Using a case study of young people, we arrived at several findings of importance for future research, concerning the dominant role of smartphone use in daily life, where such use occurs, and the extent to which many everyday doings are intervened by parallel background digital activity. We identified how young people find various ways to manage the opportunities and frictions arising from this online background and examined how the interaction between foreground and background activity is experienced and negotiated.

Acknowledgements

This chapter is based on research supported by the Swedish Research Council [grant number D0106901] and Marianne and Marcus Wallenberg Foundation [grant number 2013.0164].

References

Aguliéra A, Guillot C and Rallet A (2012) Mobile ICTs and physical mobility: review and research agenda. *Transportation Research A* 46: 664–672.

Bayer JB, Ellison NB, Schoenebeck SY and Falk EB (2016) Sharing the small moments: ephemeral social interaction on Snapchat. *Information, Communication & Society* 19(7): 956–977.

Bertel TF and Ling R (2016) 'It's just not that exciting anymore': the changing centrality of SMS in the everyday lives of young Danes. *New Media & Society* 18(7): 1293–1309.

Bertel TF and Stald G (2013) From SMS to SNS: the use of the internet on the mobile phone among young Danes. In Cumiskey K and Hjorth L (eds) *Mobile Media Practices, Presence and Politics: The Challenge of Being Seamlessly Mobile*. New York: Routledge, 198–213.

Bittman N, Brown JE and Wajcman J (2009) The mobile phone, perpetual contact and time pressure. *Work, Employment and Society* 23(4): 637–691.

Ellegård K (2000) A time-geographical approach to the study of everyday life of individuals: a challenge of complexity *GeoJournal* 48: 167–175.

Hägerstrand T (1970) What about people in regional science? *Papers of the Regional Science Association* 1(24): 7–21.

Hägerstrand, T (1985) Time-geography: focus on the corporeality of man, society and environment. In: *The Science and Praxis of Complexity*. Tokyo: United Nations University, 193–216.

Hubers C, Dijst M and Schwanen T (2008) ICT and temporal fragmentation of activities: an analytical framework and initial empirical findings. *Tijdschrift voor Economische en Sociale Geografie* 99(5): 528–546.

Hubers C, Dijst M and Schwanen T (2015) The fragmented worker? ICTs, coping strategies and gender differences in the temporal and spatial fragmentation of paid labour, *Time & Society* 27(1): 92–130.

Ito M and Okabe D (2005) Technosocial situations: emergent structurings of mobile email use. In: Ito M, Okabe D and Matsuda M (eds) *Personal, Portable, Pedestrian: Mobile Phones in Japanese Life*. Cambridge, MA: MIT Press, 165–182.

Janco RA and Cotton SR (2012) No A 4 U: the relationship between multitasking and academic performance. *Computers & Education* 59: 505–514.

Judd T (2013) Making sense of multitasking: key behaviours. *Computers & Education* 63: 358–367.

Kasesniemi E-L and Rautiainen P (2004) Mobile culture of children and teenagers in Finland. In: Katz JE and Aakhus M (eds) *Perpetual Contact: Mobile Communication, Private Talk, Public Performance*. Cambridge, UK: Cambridge University Press, 170–192.

Katz JE and Crocker ET (2015) Selfies and photo messaging as visual conversation: reports from the United States, United Kingdom and China. *International Journal of Communication* 9: 1861–1872.

Kenyon S and Lyons G (2007) Introducing multitasking to the study of travel and ICT: examining its extent and assessing its potential importance. *Transportation Research Part A* 41: 161–175.

Kwan M-P (2002) Time, information technologies, and the geographies of everyday life. *Urban Geography* 23(5): 471–482.

Lenntorp B (2005) Path, prism, project, pocket of local order: an introduction. *Geografiska Annaler: Series B, Human Geography* 86(4): 223–226.

Licoppe C (2004) Connected presence: the emergence of a new repertoire for managing social relationships in a changing communications technoscape. *Environment and Planning: Society and Space* 22(1): 135–156.

Ling R (2004) *The Mobile Connection*. San Francisco, CA: Morgan Kaufmann.

Ling R (2013) *Taken for Grantedness: The Embedding of Mobile Communication into Society*. London: MIT Press.

Ling R (2017) The social dynamics of mobile group messaging. *Annals of the International Communication Association* 41(3–4): 242–249.

Ling R and Lai C-H (2016) Microcoordination 2.0: social coordination in the age of smartphones and messaging apps. *Journal of Communication* 66(5): 834–856.

Ling R and Yttri B (2002) Hyper-coordination via mobile phones in Norway. In: Katz JE and Aakhus M (eds) *Perpetual Contact: Mobile Communication, Private Talk, Public Performance*. Cambridge, UK: Cambridge University Press, 139–169.

Lyons G and Urry J (2005) Travel time use in the information age. *Transportation Research A* 39: 257–276.

Oksman V and Turtiainen J (2004) Mobile communication as a social stage: meanings of mobile communication in everyday life among teenagers in Finland. *New Media & Society* 6(3): 319–339.

Palm J and Ellegård K (2011) Visualizing energy consumption activities as a tool for developing effective policy *International Journal of Consumer Studies*, 35(2): 171–179.

Rettie R (2009) SMS: exploiting the interactional characteristics of near-synchrony. *Information, Communication & Society* 12(8): 1131–1148.

Robinson JP and Godbey G (1997) *Time for Life: The Surprising Ways Americans Use Their Time*. University Park, PA: Pennsylvania State University Press.

Schwanen T and Kwan M-P (2008) The internet, mobile phones and space-time constraints. *Geoforum* 39(3): 1363–1377.

Schwanen T, Dijst M and Kwan M-P (2008) ICTs and the decoupling of everyday activities, space and time: introduction. *Tijdschrift voor Economische en Sociale Geografie* 99(5): 519–527.

Thulin E (2017) Always on my mind: How smartphones are transforming social contact among young Swedes. *Young* 26(5): 1–19.

Thulin E and Vilhelmson B (2007) Mobiles everywhere: youth, the mobile phone and changes in everyday practice. *Young* 15(3): 235–253.

Thulin E and Vilhelmson B (2009) Mobiles and the transforming everyday practice of urban youth. In: Ling, R and Campbell, SW (eds) *The Reconstruction of Space and Time: Mobile Communication Practices*. London: Transaction.

Thulin E, Vilhelmson B and Schwanen T (forthcoming) Absent friends? Smartphones, mediated presence and the recoupling of online social contact in everyday life.

Urry J (2000) *Sociology Beyond Societies: Mobilities for the Twenty-First Century.* London: Routledge.

Wajcman J (2015) *Pressed for Time: The Acceleration of Life in Digital Capitalism.* Chicago, IL: University of Chicago Press.

Wellman B (2001) Physical place and cyberspace: the rise of personalized networking. *International Journal of Urban and Regional Research* 25(2): 227–252.

Vilhelmson B, Thulin E and Fahlén D (2011) ICTs and activities on the move? People's use of time while traveling by public transportation. In: Brunn, SD (ed.) *Engineering Earth: The Impacts of Megaengineering Projects.* Dordrecht, Netherlands: Springer Science and Business Media, 145–154.

Vilhelmson B, Thulin E and Elldér E (2017) Where does time spent on the internet come from? Tracing the influence of information and communications technology use on daily activities. *Information, Communication & Society* 20(2): 250–263.

Yin L, Shaw SL and Yu H (2011) Potential effects of ICT on face-to-face meeting opportunities: a GIS-based time-geographic approach. *Journal of Transport Geography* 19: 422–433.

6 A relational interpretation of time-geography

Martin Dijst

Introduction

Almost half a century ago Torsten Hägerstrand presented his seminal paper "What about people in regional science?" at the European Congress of the Regional Science Association in Copenhagen. According to Google Scholar (accessed August 19, 2017), the paper has been cited an impressive 3,772 times since its publication in 1970. The growing significance of the paper is demonstrated by an enormous increase in the citation score, which was 1,772 on April 1, 2012 (Shaw 2012). Hägerstrand's human ecological approach focused on the physical world comprised of all inorganic and biological existents as well as their artifacts, which sets limits for human life (Hägerstrand 1985). In various publications (e.g., Hägerstrand 1973, 1982, 1989) he broadened, deepened and refined his framework. Although other disciplines have used his ideas (Shaw 2012), human geographers, planners, geographic information scientists and transportation researchers have predominantly embraced his framework.

Despite the popularity of this approach over the years, Hägerstrand has received considerable criticism (e.g. Giddens 1984; Rose 1993; May and Thrift 2001; Dijst 2009a). Two major issues in classical time-geography are relevant to this study. First, time-geography is based on an absolute and relative interpretation of time and space. It uses a three-dimensional time-space in which individual paths or trajectories of organic and inorganic entities are situated in relation to each other. It is a representation of "a singular or uniform social time stretching over a uniform space" (May and Thrift 2001: 5). However, people actually are also simultaneously situated in a relational framework. In a relational conceptualization of time-space, a human being is understood as a network of relationships in which consciousness of temporal and spatial distances are measured in terms of processes and activities (Harvey 2007; Law 1992).

The second and related major issue concerns the neglect of the mental and emotional experiences and meanings of human beings. Although Lenntorp (1976), Mårtensson (1979) and Hägerstrand (2009) mentioned the experienced side of life, they did not develop their ideas. In classical time-geography, the human body is treated as a corporeal entity with no observable sexual, gender or racial expressions with associated meanings in interactions. Rose (1993) stated that in time-geography the "body" is reduced to a neutral vessel carrying the person along a path through time and space. Despite this lack of attention for the inner world of people Hägerstrand has never denied time-geography's relevance for understanding the organization of daily life (Gren 2001). According to Hägerstrand (1973: 75), "Feelings and opinions contain seeds for future change" (see also Hägerstrand 1985, 1995). However, he states:

> I agree, my way of thinking is admittedly reductionistic in a specific sense. . . . But one cannot talk about anything without simplifying, that is reducing reality to something smaller than it is. I want to find the bare skeleton of what could call natural situations.
>
> (Hägerstrand 1989: 2)

The question is whether Hägerstrand has eliminated too many essentials in human life in a search for the bare skeleton of natural situations. Is the material basis sufficient to understand human life? In any case, it is the fundamental basis of all interactions and might be sufficient for understanding purely instrumental tasks, such as shopping for groceries, commuting to work or visiting a bank. However, in essence all activities in time and space imply interactions with the environment, which are accompanied with mental and emotional experiences and meanings. This phenomenon can be expressed in such examples as fears while biking at night or in joyful experiences while walking in a park with your children. For time-geography to limit interactions in daily life to instrumental needs and thus neglect other dimensions of "being in the world" places a serious constraint on achieving the full potential of the time-geographical framework. This study will clarify the potential of time-geography to study experiences with and meanings of relationships between people and their environments while they are carving paths in their spatio-temporal landscapes. Such an extended version of time-geography can increase understanding of health and social issues, such as social integration and cohesion.

This study aims to develop an extended and modified conceptual time-geographical framework that meets the demands of existential and relational human beings' needs when organizing their daily lives. To that purpose, theoretical perspectives derived from, e.g., actor-network theory, (post-)phenomenology, emotional geography and assemblage theory are applied to develop new theoretical concepts that represent people's relatedness to the world. In the next section, existential thoughts and feelings will be discussed that will be elaborated in terms of relational needs in the section thereafter; and in the section that follows, the classical time-geographical framework will be extended and modified with relational (re)conceptualizations. The study will end with some conclusions and an exploration of avenues for future research.

Existential foundations of human life

Time-geography places the material basis as an existential foundation of human life at the center. However, other existential foundations will be discussed in this section. Since the emergence of *Homo sapiens* in Africa more than 200,000 years ago, humans have lived for 99% of their time in small hunter-gatherer clans with-out all the material products, technologies and cultural and economic expressions that are so familiar to many of us living in the urbanized and mobile world. Due to technological, cultural and economic developments, we were able to create our own worlds, or as Perry (2002: 80), a leading authority in child psychiatry, states: "we have invented ourselves". We are living now in a world full of oppor-tunities: "the world becomes an infinite collection of possibilities: a container filled to the brim with a countless multitude of opportunities yet to be chased or already missed" (Bauman 2000: 61). Despite these overwhelming developments, we could have the impression that we have freed ourselves from nature. As Perry (2002: 80) makes clear: "Yet we are biological creatures, bound by the laws of nature to a time-limited existence" (see also Corning 2000).

Perry (2002) emphasizes that the brain's primary mandate is survival of the species. The three objectives to accomplish this mandate include: (1) stress and threat responses; (2) mate selection, reproduction and protection; and (3) nurtur-ing dependents. The existence of these existential needs does not mean that we always live to support these needs. For example, we can refer to the dysfunctional use of food and drinks or healthcare practices that can undermine people's health (Corning 2000: 47). In that sense, these existential needs are not deterministic. To a certain extent, some of these biological needs can also be found in Hägerstrand's (1970: 12) capability constraints: "the necessity of sleeping a minimum number of hours at regular intervals and the necessity of eating, also with a rather high degree of regularity".

The most important strategy to meet Perry's three objectives is "to create relationships. Relationships which allow us to attach, affiliate, communicate and interact to promote survival, procreation and the protection of dependents" (Perry 2002: 81). Creating relationships is not limited to concrete relations with family, community members or friends to meet biological demands. From a philosophi-cal point of view, creating relationships refers to relationships with the world. Heidegger (1926/1962) argues that people are always disposed to their everyday involvement in the world but that the specific modes ("moods" or "Stimmung") through which this "Befindlichkeit" is expressed and realized differ (Guignon 2009). A mood is a state of mind that is hidden in the background, preconscious and all-pervasive. In an existential respect, a mood "implies a disclosive sub-mission to the world, out of which we can encounter something that matters to us" (Heidegger 1926/1962: 177). Ratcliffe coined the term "existential feelings", which is slightly different from Heidegger's concept of "mood" (Ratcliffe 2009a: 180): "they constitute a sense of relatedness between self and world, which shapes all experience . . . they give us a changeable sense of 'reality' and of 'belonging to the world'". In other words, moods and existential feelings allow people to engage meaningfully with the world. Although existential moods and feelings precede all concrete psychological moods, feelings or conditions, they inform specific daily life experiences, beliefs and thoughts not as separate from them but as pre-reflective and pervasive parts of life.

How can this disposition to everyday involvement in the world be understood? To answer this question, a better understanding of the anthropological foundations of human life in the world is needed for which Plessner's theory of positionality can be seen as a source of inspiration. Compared to that of plants and animals, the positionality of human beings is complex. People take an "ex-centric position" ("exzentrische Positionalität"). Plessner (1928/1975: 292) argues: "He not only lives and experiences, he also experiences his experience" ("Er lebt und erlebt nicht nur, sondern er erlebt sein Erleben"), which can be understood through an example taken from daily life (Mishare 2009: 135): "When we speak, gesture, or write, we are simultaneously recipients, witnesses, of our own communicative efforts. We hear our own voice and partially see our bodily gestures". This ex-centric positionality involves human beings continuously experiencing shifts between centered positions –being inside the body in the center of one's world (the "Inner World" or "Innenwelt") – and his ex-centered position outside his body at a distance (Fuchs 2005; Fischer 2009: 158):

> From within, he feels like a centered living subject, but at the same time, by observing himself out of the corner of his eye, at a distance, he finds himself as a body among material bodies, marginalized, de-centered, objectified, like a 'mere animal' [Plessner], a thing among other things.

The intrinsic sense of distance is essential for the relationship to the "self", which constitutes being a "person" (Rehberg 2009). This ex-centric positionality enables people to transcend the here and now and adopt the others' perspective on themselves (Fuchs 2005; Lindemann 2009). For example, a person can think about buying new clothes and imagine how they would fit him and how others would react to his new outfit. In other words, this transcending ability enables people to *reflect* on relationships with themselves and other animate or inanimate things and to *change* these according to the expectations of others (Heinze 2009; Lindemann 2009).

The ex-centric positionality of human beings and their ability to reflect essentially means that people are not biologically determined, to a large extent and certainly in comparison with plants and animals. They consequently experience an existential "homelessness" or "rootlessness" ("konstitutiven Heimatlosigkeit" or "Wurzellosigkeit") by not knowing how to live their lives in the Outer World (Plessner 1928/1975: 309). This Outer World consists of organic and inorganic entities from the person's surroundings and to which a person as body ("Körper") – that is a material object that takes up room in absolute space – is situated (as in time-geography: see Hägerstrand 1970, 1973; Dijst 2009a). One could say that people have a fragile equilibrium in which they experience being "highly vulnerable and at risk" (Heinze 2009: 126). Human beings are struggling with how to relate themselves to the world. They oscillate between their centric and ex-centric positions. This struggle is also discussed in Heidegger's *Being and Time* (1926/1962). Analogous to Plessner, Heidegger (1926/1962: 163) characterizes this struggle as follows:

there is constant care as to the way one differs from them ["Others"], whether that difference is merely one that is to be evened out, whether one's own Dasein ["Being-there"] has lagged behind the Others and wants to catch up in relationship to them, or whether one's Dasein already has some priority over them and sets out to keep them suppressed.

This struggle over how to relate oneself to the world can be differentiated for at least three existential feelings: anxiety, affection and significance. This study discusses these three existential feelings that are omnipresent and also demonstrate what matters in human existence. The unconscious experience of these existential feelings will lead to different senses of belonging to the world.

Anxiety ("Angst") is a basic existential feeling in human life, which is a central theme of Heidegger's *Being and Time*. As an existential feeling, it does not refer to a daily psychologically experienced "fear" of something specific but to a person's general latent worrying about their involvement in and their belonging to the world. "In anxiety one feels 'uncanny'" (Heidegger 1926/1962: 233), which actually means that people feel "strange", "mysterious" or "not at home" due to their disconnections from ordinary daily life and their orientations in life (Merleau-Ponty 1945/2005). Heidegger sees this mood as something positive in the sense that it returns the initiative for "Being-in-the-world" to the individual person. It discloses a person's possibilities for changing his relations to his environment. However, the existential feeling and related reflections might also lead to a suppression of this anxiety and a continuation of the existing life.

Although Heidegger acknowledges the existence of moods other than anxiety, he ignores these in his discussion in *Being and Time*. Ratcliffe (2009a, 2009b) tries to close this gap in Heidegger's philosophy to achieve a better match between a person's existential orientations and his daily feelings and embodied experiences. Moreover, through anxiety, existential feelings can be expressed in terms such as "emptiness", "estranged", "alive", "being at home", "loneliness", "separated", "insignificance", "settled", "harmony" and "belonging" (Ratcliffe 2009a). A long list of existential feelings seem to exist.

Affection, the second existential feeling, is often associated with love and can be seen as a close, intimate personal attachment to another person, which can be distinguished from other less "in-depth" modes of concern or caring attached to animals and inanimate objects. The subjects of affection are not the distinctive characteristics of a person's behavior or attitudes but instead the broader concern of a person's living or his personhood (Helm 2009). Velleman (1999) places this interpretation of love against the backdrop of Kant's philosophy, in which a person's dignity demands that we value people as individuals and not as members of a group. Comparable with but different from anxiety, affection also has an affordance to disclosure. According to Velleman (1999: 361), "love is an exercise in 'really looking'. . . . Many of our defences against being emotionally affected by another person are ways of not seeing what is most affecting about him [the Other]". He argues that this opening-up is in the interest of the other person and not of us. Helm (2009: 61) also emphasizes this idea:

> in identifying with another person in the way just outlined, one is not con-
> cerned with one's own well-being but rather with his [the other person's].
> This is to be committed to his identity as this person for his sake (and not for
> any ulterior motive).

These statements inappropriately conflict with the existential interpretation of
reflexivity formulated by Plessner as well as Heidegger, which demonstrates that
a person's ex-centric positionality and existential feelings always concern the
relatedness of the person to the world. This concept means that the affection for
another person is assessed in terms of the meaning of this affection for a person's
belonging to the world.

The third existential feeling to be elaborated upon is *significance*. This feeling
refers to having a great effect or being of high value in the world of the person.
According to Brogaard and Smith (2005: 443),

> a meaningful life is a life upon which some sort of valuable pattern has been
> imposed – a pattern which relates not merely to what goes on inside the per-
> son's head, but which involves also, in serious ways, the person having an
> effect upon the world.

James (2005) sees a life as meaningful when a person has achieved something.
The realization of such a life must be the result of the efforts and decisions of
oneself in a particular situation and not at the behest of others (Brogaard and
Smith 2005). In essence, this refers to the subject character of human existence
expressed in a person's constant care to observe his or her difference from others. As such, a person can take control of his or her existence and decisions and,
instead of living an "average" life similar to others, live an "authentic" ("eigen-
tliches") life (Heidegger 1926/1962: 165–167). Each person's significance is
assessed against the relevant public standards (Brogaard and Smith 2005), which
are related to the diversity of groups or communities to which a person belongs.
Similarly to anxiety and affection, significance discloses possibilities for change
in one's existential relationship with the world.

The three existential feelings each have a dual character. As argued by
Heidegger, anxiety can offer people opportunities to change their life orientations.
However, it can also be experienced by people as unpleasant, a feeling that might
be interpreted as to avoid. Similarly, affection is two-faced, as Velleman argues
(1999: 361): "Love disarms our emotional defences; it makes us vulnerable to
the other". What is more significant, it can constitute a state of belonging to the
world but it can also lead to a suppressive impact by others. A person must find a
balance in his orientations towards these existential feelings. This existential balancing also applies to the intimate relationships between the existential feelings.
Anxiety can increase awareness of the meaning of affection and significance, and
this effect can support these existential feelings but could also harm them when-
ever anxiety is overwhelming. Conversely the loss of affection or significance can
stimulate anxiety and feelings of being disconnected from life. In addition, affection and significance are intertwined. In close, intimate relationships a person

can also experience having a special effect on the other. Being significant can, in itself, stimulate the existential feeling of affection.

Relating oneself to the world in terms of anxiety, affection and significance is accompanied by stress. Stress processing is at the base of all existential feelings. It can lead to positive stress ("eustress") such as experiencing adrenaline while making an exceptional sport achievement or falling in love with somebody. It can also lead to "distress" when people feel alienated, lonely, not at home, not seen, or not experiencing affection from others. Problems arise when the stress response does not switch off (Selye 1977; Joye and Van den Berg 2011). Stress-hormone levels that stay too high for too long cause high blood pressure and suppress the immune system. A strong relationship exists between stress and cancer, heart disease and psychiatric disease (Rice 2012). Perry (2009: 246) argues that creating relationships is the only way to address distress and survive in the natural world: "the presence of familiar people projecting the social-emotional cues of acceptance, compassion, caring, and safety calms the stress response of the individual: 'You are one of us, you are welcome, you are safe'". This idea particularly applies to the relationship between caregivers and a child at home or relationships with family, friends, communities and the public in later stages in life and other social situations that afford reduction of stress, feeling connected and belonging (Dijst 2014).

Existential feelings inform relational needs

Anxiety, affection and significance are fundamental existential feelings that connect people to the world but precede will or consciousness. Nevertheless, these feelings shape all experience (Ratcliffe 2009a) and thus inform relational needs of human beings (Baumeister and Leary 1995; Kumashiro et al. 2008; Erskine 2010). According to Stewart (2010): "Relational needs emerge out of our social connectedness, and help sustain and nurture our emotional attachments to others." To achieve a better match with time-geography and the meaning of relations in concrete daily life experiences, we translate the existential feelings discussed in this study into practical relational needs.

Based on his long-standing experience as a psychotherapist, Erskine (Erskine and Trautmann 1996; Erskine 1998; Stewart 2010) derived a set of eight relational needs important in relationships with other people in daily life. Although originally developed for ("significant") relationships with relatives, friends and/ or colleagues, these needs might also be effective and significant in the fluid, frequent relationships with less well-known, or even anonymous, others we encounter on a "daily" basis in public places or while on the move. Relational needs are neither defined nor "owned" by an individual person but, as the nexus of relations, are mutually constituted by what is "inside" a person and how they are related to others (Slife 2004). When all or some of these relational needs are unmet, the result might be discomfort manifested in frustration, anger or aggression (Erskine 1998). An accumulation of these experiences of psychological distress may develop into disturbances in mood, thought, physiological processes and behavior (Stewart 2010).

Figure 6.1 presents the existential feelings anxiety, affection and significance in the background, displaying at the forefront relational needs informed by these fundamental feelings, which inspire our relationships and feed our attachments. As an existential feeling, *anxiety* largely affects the need to experience "security". According to Slife and Wiggins (2009:20): "Fear of rejection – the fear that we do not belong, are not acceptable, or do not have meaningful relations – is the greatest of all the fears and anxieties." Protecting physical and emotional vulnerabilities is important to offer a person the opportunity to be open and vulnerable without ridicule, humiliation, shame, blame or verbal or physical violence. This situation is usually the case among good friends at home or while attending public events together. In these situations, one can feel at home (Wilton 1998; Duyvendak 2011). Violation of this relational need could mean that a person is (partly) closing off himself in relationships with others and avoiding certain people or situations. An example is avoidance of night-life districts that have a questionable reputation as "violent areas" (Brands 2014).

Figure 6.1 Existential feelings and relational needs

Affection informs several closely related relational needs such as experiencing "care and love", "acceptance", "mutuality" and "initiatives from others". These relational needs can be fulfilled by opening up oneself, such as presenting a gift or doing something nice for another person and experiencing affection, gratitude and feelings of acceptance and being valued. Special cases in this respect are relationships with parents, teachers, mentors and others on whom a person relies for protection, encouragement and advice (Stewart 2010). We expect that representatives in public life, like policemen, receptionists, doctors, bus drivers, shopkeepers or others with public functions, will support us when approached with respect. Taking initiative excessively without experiencing "initiatives from others" in such cases as making appointments with social network members or displaying examples of good behavior in public (e.g., giving up your seat on a crowded bus for an elderly person) can be frustrating and dissatisfying. This need is related to the need to experience meaningful, loving and impact-rich relationships.

"Mutuality" refers to the presence of others who are similar to a person and can understand experiences without explanation, which might be the case in romantic relationships. In daily life this need can be experienced by wearing clothes that are "accepted" as being appropriate by others. This phenomenon can be illustrated by school children who influence each other by wearing clothes that represent their "lifestyle". Slife and Wiggins (2009: 22) highlight the reverse of mutuality. Striving excessively to find similarities might lead to inauthentic and manufactured "sameness". In these relationships, a person assumes the other's attitudes without providing true closeness, intimacy or belonging.

Significance shapes related needs, such as being "meaningful" and "unique" and having an "impact". To experience validation, recognition, understanding and acceptance for what a person is doing, saying or feeling, or who he is are at the sources of these relational needs. To fulfill this need, the person needs to be open and reveal himself to others, initiating contacts. The reversal of this active positioning towards others is the risk of being rejected or of not being accepted by others as a person who is valuable or who is acting or reacting in a worthwhile manner. Such a situation can lead to resentful or wounded experiences and might harm another relational need: "security".

Experiencing influence on other person(s) means having an acknowledged effect on, or sometimes power over, others to change their thoughts and behavior or to provoke an emotional reaction from them in a desired way. This need can be an expression of the relational need to be unique. Having a discussion with others or accompanying an elderly person across a busy street are examples of this need. The experience of having an impact can be related to fulfillment of immediate hedonistic emotional satisfactions or realizing higher ethical or spiritual values such as raising your children and supporting others, which might be accompanied by (some) sacrificing or even suffering (Slife and Richardson 2008). As such, this relational need can be linked to affection-related needs.

"Uniqueness" can refer to one's identity. In contrast to the need for "mutuality", a person can also feel the need to communicate "one's self-chosen identity through the expression of preferences, interests and ideas without humiliation or rejection" (Erskine 1998: 240). Our relational "self" actually has a dialectical quality: our identity stands out when others are more similar than ourselves (Slife and Wiggins 2009) as expressed clearly by Slife (2004: 166): "I am who I am, in part, because of who you are". As previously emphasized by Erskine's quote, uniqueness can lead to humiliation in some situations, particularly when one's identity is not self-chosen, as evidenced by the experiences of a poor single mother on a San Francisco bus who said:

> When you don't have a car and you go grocery shoppin' and you're strugglin' with the bags on the bus. Oh, it's so embarrassing! And everybody else is on the bus and you're trying to grab your stuff and get on the bus and they're lookin' at you.
>
> (McQuoid and Dijst 2012)

Although relational needs were originally focused only on relationships between people, these can also be extended to non-human entities. Latour (2005) stated that we have relationships with not only other people but also other organisms and inanimate entities (see also Slife and Wiggins 2009). People can meet relational needs with pets, such as dogs and cats, with which they can share care, or even love, or influence (Fox 2006). Inanimate objects can also be involved in relational networks. For example, consider the embodied nature and emotional attachments of clothing (Longhurst 2001) by which the experience of the body as well as the presentation of the body or the self can change (Colls 2004: 587). In this manner, clothes can respond to relational needs such as security or feelings of uniqueness. In a different manner, relational needs can be present in inherited furniture, jewels or other belongings that carry precious memories of beloved deceased ones. In the presence of the beneficiary, these attributes can offer the person comfort and feelings of contact with the late relatives or friends (Parker and Dunn 2011). The implications of identification of existential feelings and relational needs for time-geography will be discussed in the next section.

A relational (re)conceptualization of time-geography

This study has already discussed both the material basis of life central to time-geography and other existential foundations of human life. These foundations are related to biological demands such as stress regulation, protecting and caring for offspring and sensing connectedness and attachments to the world. Inspired by this discussion, this section will develop a relational conceptual framework suitable to address experiences of existential feelings and relational needs in daily interactions in geographical environments. The conceptual framework of classical time-geography remains at the heart of this approach.

Figure 6.2 presents the time-space of time-geography. In this 3D representation of reality, individuals (A, B, and C) and other organic and inorganic entities (such as the vertical path of solid waste and, depending on wind direction, vertical or diagonal path of car pollution) and natural phenomena, such as sunrise (dark zones in Figure 6.2) and weather (e.g. diagonal path of a rain cloud), describe uninterrupted "paths" through time and across space (Hägerstrand 1970; Ellegård and Svedin 2012; Dijst 2009a; Kwan 1998; Miller 2005). The timescale of these paths for human beings can vary from a day to the whole human life span. Consistent with Galileo's observation that "A thing's place was no longer anything but a point in its movement, just as the stability of a thing was only its movement indefinitely slowed down" (Foucault 1986: 23; Dijst 2013), every organism or thing is constantly in motion, even when they are themselves at rest. Individual paths come together in two types of *encounter situations*: "stationary", when individuals and other entities and natural phenomena are present in activity places such as a dwelling, an office, a shop or a park where all participate in activities (in Figure 6.2: the vertical paths of person B and C come together); and "mobile", when people or mobile objects and natural phenomena are moving from one place to another, participating in activities while moving (in Figure 6.2: the diagonal paths of person A and B come together). Both situations differ in stability, but they actually are always dynamic in nature to various degrees. For example, when a person is stationary in a certain activity place, other people, mobile objects and natural phenomena may move in and out the person's range of sensory perception. Similarly, landscapes shift and alternate as an individual moving in space changes position over time.

In classical time-geography, these encounter situations are treated as networks composed of uninterrupted paths and labeled as "bundles" (Hägerstrand 1970; Dijst 2009a). However, the interactions between the actors from whom

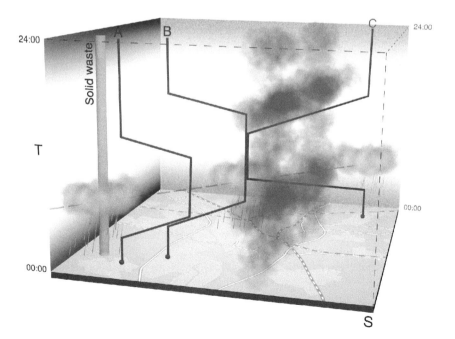

Figure 6.2 Time-space paths of organic and inorganic entities

these paths originate are not taken into account. This concept was also a criticism Giddens (1984) made of time-geography and the introduction of the concept "locale" as a setting of interaction in his structuration theory. The interaction process between actors is also central to the actor-network theory (ANT) developed by Latour (1999, 2005), Callon (1999) and Law (1993): "Modern societies cannot be described without recognising them as having a fibrous, threadlike, wiry, stringy, ropy, capillary character that is never captured by the notions of levels, layers, territories, spheres, categories, structure, systems" (Latour 1997: 2). Understanding how various material and non-material entities become related, act as a network and change over time is the major objective of ANT. However, ANT does not specifically explain these relationships between actors (Dijst and Gimmler 2016), which is an important obstacle to enriching time-geography with the ability to grasp other existential foundations and relational needs in human life, as discussed in the sections above. In that respect, an embodied ontology, which takes the physical body as the starting point for experiencing and being in the world (Hubbard 2005; Davidson and Milligan 2004; McQuoid and Dijst 2012), is indispensable.

From this perspective, in time-geography, relations between actors (Figure 6.3: A–D) can be captured by introducing the concept of *relational string*, which is a threadlike connection between actors present in their spatio-temporal paths and which can last for a moment or continue for the time the paths are together. Consistent with Raunig (in Dewsbury 2011: 150), the interaction or exchange between actors occurring in a relational string can be seen as "(an) unbounded flow between bodies that touch or come close to one another". A string can move or vibrate in different ways. In a string, impressions, perceptions, attitudes, ideas, emotions and other types of information emerge and can be transmitted in different rhythms and intensities. This information originates from bodily or corporeal appearances, such as biological traits (e.g., race, gender, age, weight) and cultural or lifestyle expressions (e.g., behaviors; wearing clothes, glasses, jewelry; using crutches) that have social meanings. This embodied reinterpretation of time-geography addresses Rose's (1993) "neutral body" criticism of time-geography.

A relational string does not stand independently but is part of an *assemblage of relational strings*. An assemblage represents different types of associations of relations that may be temporary, such as fluid contacts with people, objects or natural phenomena or longer lasting contacts, such as with family members or friends. Because assemblages are constantly opening up to new relations and new becomings, they are never finished (Anderson and McFarlane 2011; Lancione 2013; Dewsbury 2011). Although they resemble "actor-networks" in ANT (Latour 1999; Murdoch 1998), assemblages are slightly different in the sense that properties of entities emerge in relationships rather than pre-given as in ANT (Anderson and McFarlane 2011). In assemblages, actors are not seen as discrete categories but are continuously entangled, co-constituting and co-affecting (Lancione 2013). From an actor perspective, actors are part of or "are (becoming) assemblages: they are composed of flesh and bones, thoughts and wishes, which relate, change and move" (Lancione 2013: 359). Law (1992: 3)

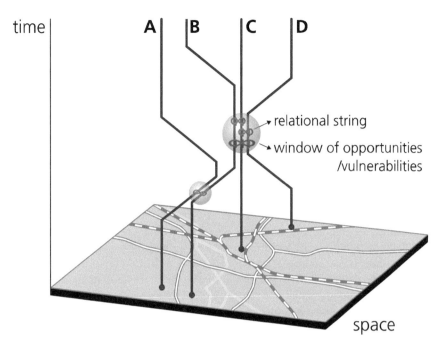

Figure 6.3 Relational strings in the time-space

convincingly specifies how his identity changes as a consequence of changes in his assemblage relations: "If you took away my computer, my colleagues, my office, my books, my desk, my telephone, I wouldn't be a sociologist writing papers, delivering lectures, and producing 'knowledge'. I'd be something quite other – and the same is true for all of us."

In these relational strings and assemblages, people constantly experience *embodied exposure* to other entities and phenomena in their environment. This term was initially coined by Levinas (Harrison 2008; Lok 2011) and refers to the experience of the immediacy of bodily relationships (Lok 2011: 113 and 147). This experience triggers a bodily sensation prior to will or consciousness and makes it possible to sense or comprehend the world and take responsibility for it (Lok 2011: 125). A part of this embodied exposure is the empathic ability of people to sense the weakness, vulnerability and hunger of others and the ability to sense the pain of others because their lives are intertwined (Critchley 2006; Lok 2011). Embodied exposure also refers to other embodied emotions such as feelings of pleasure, love, humiliation, fear, pride, care and loneliness that are intricately connected to relational needs and existential feelings. Embodied experiencing extends beyond sensing the presence of other entities and phenomena in the environment. We know from Plessner (1928/1975) that due to ex-centric positionality, people are also experiencing and reflecting on themselves, which is experienced as an existential struggle.

Embodied exposures are dependent on the spatio-temporal situations in which people are assembled with other people, entities and natural phenomena. From a time-geographical perspective, these situations are relevant in three respects. First, these situations refer to the absolute position in time and space of actors as expressed in Figures 6.2 and 6.3. At various moments in a day or locations in space the presence of other people, things or natural processes might be different. Second, situations and relational strings differ in duration (Figure 6.3). One can hypothesize that the more time people spend in an assemblage of relational strings, the more they experience embodied exposure to these relational strings and the more these co-constitute and co-affect each other (see also Dewsbury 2011: 150). In general, these exposures and accompanying experiences are much longer in duration at work than while walking from the bus stop to the railway station. Finally, the path concept clarifies that the accumulation of embodied exposures acquired in various spatio-temporal situations over the day (or even life course) could affect the meaning of embodied exposures. For a person traveling daily between 08:00 and 09:00 by train to work, the train journey might be a more negative experience than for somebody who travels on the same train for the first time (Bissell 2010).

From the perspective of feelings involved in or belonging to the world, these spatio-temporal situations are relevant in two respects. First, the spatio-temporal situation and assemblages of relational strings feelings of affection and significance can emerge in *windows of opportunities to belong* (the spheres in Figure 6.3), which intensifies belongingness. For example, this situation is the case when people assist others in crossing a busy street. Second, spatio-temporal situations might also function as *windows of vulnerability in belonging* (the spheres in Figure 6.3). In the assemblages of relational strings, people might feel neglected by others and experience anxiety. Negative stress often accompanies these embodied experiences. These three-dimensional windows are two sides of the same coin because extreme or insidious neglect can lead to a decrease in opportunities to belong, and opportunities to belong can reduce experiences of vulnerability. These two types of windows are different from the "prism" in time-geography. A time window is a block of time delineated by coupling constraints of fixed activities at base locations in which travel and non-base location activities can be conducted (Dijst 2009a). A window of opportunity or vulnerability might be a part of this time window while traveling from one base location to another, but its relations with other entities are not limited to physical opportunities offered by the time window and the linked prism. People can be in contact with others by phone or internet who cannot be reached physically in the time window. They can also feel related to others or events by recalling without contacting them. Finally, the windows of opportunity or vulnerability are not limited to time windows but also apply to base locations at which people spent time.

A relational conceptualization of time-geography can be developed by introducing the discussed concepts of relational string, assemblage of relational strings, embodied exposure, windows of opportunity to belong and windows of vulnerability in belonging. However, a relational reinterpretation of existing time-geographical concepts can also be made. One of these concepts is "authority constraint". This constraint originally referred to external authorities that regulate access to activity places (Dijst 2009a). McQuoid and Dijst (2012) have reconceptualized "the authority" and extended its range. Both "authorized others" and the individual can be an authority who can impose constraints on the contact with other people or entities in the environment. These constraints include perceived or presumed appreciation and acceptance of an individual's biological and cultural appearance and behavior as a potential limitation to an individual's presence in certain spatio-temporal situations.

Not only "authority" but also "capability" and "coupling" constraints carry relational meanings. Mobile phones offer the capability to contact others who are not physically present, but the phones can also express identity similar to face-to-face communication via their brand, ringtone and color (Dijst 2009b). Visual impairment can significantly limit people in navigating in a city, and it also often carries social perceptions of others such as stares, pity and hostility, which can harm spatial competences of the impaired (Worth 2013). Especially, coupling constraints are important when discussing relational strings. The coming together of paths is essential for developing a relational string. Schwanen (2007) has demonstrated that the absence of a cuddle toy for a child in a nursery can become an even more important coupling constraint than a simple mismatch in space of the child and the toy. The toy becomes loaded with anxiety and guilt for the mother when she realizes the child depends on it to fall asleep.

As mentioned earlier human beings experience an existential struggle about how to relate themselves to the world. In principle, two ways to manage related stress and relational needs exist. First, people can try to "position" their time-space paths to be able to enter spatio-temporal situations in which their relational needs will be fulfilled and stress levels are acceptable, due to the presence of preferred others, things and natural phenomena. At the same time, they can try to avoid spatio-temporal situations in which this outcome is less feasible. For example, a person could walk her dog on a sunny day in a park out of the need to care and influence someone. Similarly, a person could avoid high-crime areas at night to reduce his vulnerability. People may choose to live in a gated community to protect themselves against relations with unfamiliar or undesirable others.

Of course, the choices available to people to manage their relationships and relational needs are dependent on their goal-directed activities such as going to work, buying groceries and accompanying children to school and the effects of "capability", "coupling" and "authority constraints". However, as demonstrated, the reverse is also possible: instrumental behavioral choices can be dependent on the preferences and constraints derived from relational needs. These examples clarify that people's daily and life paths and relational strings are intertwined.

The second way to manage relational needs and stress in relationships is by "enveloping" oneself with an open and moldable boundary sphere (Hoggett 1992). According to Plessner (1928/1975), these boundaries embrace a double-positioning of organisms to their own bodies and those of other(s) in the environment. The meaning of this double-positioning is expressed clearly as follows (Lindemann 2010: 279):

> The living body uses its boundary to close itself off from its surroundings, to make itself into its own self-organizing domain. At the same time, the living body relates to its surroundings by means of its boundary. This boundary allows it to independently enter into contact with its surroundings.

In managing the "information flows" in their relationships to meet their relational needs, humans experience "stomatal oscillations" (Steppe et al. 2006), opening and closing themselves to their environments. For example, people are wearing clothes that cause them to look tough to protect their vulnerability, or they listen to music on their mobile phone via headphones to feel comfortable. By wearing appropriate clothes and being encouraged by friends, people may also go wild, dancing and drinking alcohol (McQuoid and Dijst 2012).

Conclusions and discussion

This study began with the premise that Hägerstrand's choice to limit his time-geographical framework to the material basis of human life has constrained the use of his ideas beyond instrumental needs and behaviors within human geography, particularly in other disciplines such as psychology, sociology and public health. In addition, this reductionist perspective is unjust to the richness of human life and prevents a more comprehensive understanding of human experiences and behavioral decisions. As this study demonstrates, an extended time-geographical framework which includes existential feelings, relational needs and stress regulation can open new avenues for analyzing social and health related issues.

With urbanization levels rising to 70% by 2050, cities will be experienced as the natural habitat of people. Living in cities extends beyond simply having a place of residence. It is also accompanied by increasing mobility, extended personal social networks and exposure to fragmented, temporary, messy and sometimes intense "fluid" contacts with unfamiliar people, animate and inanimate entities in public life (Wirth 1938; Wittel 2001; Wellman 2001; Sheller and Urry 2003). Increased mobility and activities expose people to changing and dynamic environments along their daily paths through time and across space in which they embodiedly experience relations with other entities and natural phenomena. In each of these relational strings, perceptions of the gender, skin color, dress-style, sexuality, behaviors and related emotions and attitudes of others are transmitted and may combine with other relational strings to form the basis for judgments of others and socially undesirable behavior (Ahmed 2000; Puwar 2004; McQuoid and Dijst 2012). An extended model of time-geography offers the opportunity to understand the meaning of the individual and accumulated spatio-temporal situations in both daily life and the life course for developing judgments and prejudices that influence social interactions and social cohesion. In this model, these meanings are related to anxiety, affection, significance and other existential feelings of belonging, which inform relational needs. One of the issues is to explain why spatio-temporal situations differ in affordance to the development of negative judgments and avoidance of interactions.

Social interactions in cities often increase negative stress due to feelings of annoyance, insecurity or loneliness (Abbott 2012). These stressful experiences might lead to mood and anxiety disorders, depression and schizophrenia, which are markedly more common in people raised in cities (Van Os et al. 2010; Lederbogen et al. 2011). Physical medical problems, such as cardiovascular disease and diabetes and specific types of cancer are related to these stress experiences (Abbott 2012; Blackburn and Epel 2012). Public health and epidemiology researchers increasingly emphasize the relevance of daily and lifelong environmental exposures for health and health behaviors (Wild 2012; Perchoux et al. 2013; Merlo 2012). Several attempts to apply the framework on health studies to time-geography have occurred. Inspired by Lenntorp, Schærström (1999; Aase 1997) noted the relevance of measuring exposure to environments along life paths to identify associations and causal relationships between individuals and the occurrence of ALS and leukemia. Sunnqvist et al. (2007, 2013; Örmon et al. 2015) have demonstrated the relevance for psychiatric diagnoses, care and treatment of reconstructing and space-time mapping of household moves and major social events (e.g., loss of parents, divorce, birth of a child, change of job) over the life course. In the domain of occupational therapy, the time-geographical approach has been applied to enrich therapeutic practices (e.g. Orban et al. 2012; Bredland et al. 2015). In particular, the positioning in time and space of stress events (e.g., problems at work, interpersonal problems, blows to self-esteem) and the use of maladaptive coping strategies to cope with stressors have strengthened a comprehensive understanding of a person's life history.

Although valuable results for health research have been documented with the classical model of time-geography, the extended model, based on existential feelings and relational needs, can lead to a more thorough understanding of the implications of environmental exposures for public health. First, in addition to the life course, embodied exposures in daily life are important to understand health and health behaviors as well. In all daily spatio-temporal situations, people are exposed to health risk factors in the social and physical environments (McQuoid and Dijst 2012; Richardson et al. 2013). In particular, the interactions between residential and daily mobility and their health implications are hardly understood. Second, the framework can pinpoint exactly which features of the social, built and natural environments, such as crowds and lack of green spaces, cause psychological distress and which coping strategies or emotion regulatory processes (Gross 1998) are most effective for managing stress while meeting relational needs. Related issues include why vulnerability experienced in spatio-temporal situations differs in relational needs whereas in other cases people experience belonging and which spatio-temporal positioning strategies and boundary spheres people use.

In these and similar social issues, a relational interpretation of time-geography demonstrates the importance of uniting the material with the relational existential foundations of human life. As such, individual human beings are no longer isolated from their mental and emotional inner worlds in their situational settings. This unifying perspective enriches the time-geography of its founding father Hägerstrand.

References

Aase, A. 1997. Pathogenic paths? A time-geographical approach in medical geography. *Geografiska Annaler*, 79B: 57–59 (review).

Abbott, A. 2012. Urban decay: scientists are testing the idea that the stress of modern city life is a breeding ground for psychosis. *Nature* 490: 162–164.

Ahmed, S. 2000, *Strange Encounters: Embodied Others in Post-coloniality*. London: Routledge.

Anderson, B., and C. McFarlane. 2011. Assemblage and geography. *Area* 43: 124–127.

Bauman, Z. 2000. *Liquid Modernity*. Cambridge: Polity Press.

Baumeister, R.F., and M.R. Leary. 1995. The need to belong: desire for interpersonal attachments as a fundamental human motivation. *Psychological Bulletin* 117: 497–529.

Bissell, D. 2010. Passenger mobilities: affective atmosphere and the sociality of public transport. *Environment and Planning D* 28: 270–289.

Blackburn, E.H., and E.S. Epel. 2012. Too toxic to ignore: a stark warning about the societal costs of stress comes from links between shortened telomeres, chronic stress and disease. *Nature* 490: 169–171.

Brands, J. 2014. *Safety, Surveillance and Policing in the Night-Time Economy: A Visitor Perspective*. Utrecht: Utrecht University (PhD thesis).

Bredland, E.L., E. Magnus, and K. Vik. 2015. Physical activity patterns in older men. *Physical & Occupational Therapy in Geriatrics* 33: 87–102.

Brogaard, B., and B. Smith. 2005. On luck, responsibility and the meaning of life. *Philosophical Papers* 34: 443–458.

Callon, M. 1999. Actor-network theory: the market test. In *Actor network theory and after*, ed. J. Law and J. Hassard, 181–195. Oxford: Blackwell.

Colls, R. 2004. Looking alright, feeling alright: emotions, sizing and the geographies of women's experiences of clothing consumption. *Social & Cultural Geography* 5: 583–596.

Corning, P.A. 2000. Biological adaptation in human societies: a "basic needs" approach. *Journal of Bioeconomics* 2: 41–86.

Critchley, S. 2006. Introduction. In *The Cambridge Companion to Levinas*, ed. S. Critchley and R. Bernasconi. Cambridge: Cambridge University Press. Available at: http://dx.doi.org/10.1017/CCOL0521662060 (accessed December 28, 2015).

Davidson, J., and C. Milligan. 2004. Embodying emotion sensing space: introducing emotional geographies. *Social & Cultural Geography* 5: 523–532.

Dewsbury, J.-D. 2011. The Deleuze-Guattarian assemblage: plastic habits. *Area* 43: 148–153.

Dijst, M.J. 2009a. Time-geographical analysis. In *International Encyclopedia of Human Geography*, ed. R. Kitchin and N. Thrift, 266–278. Oxford: Elsevier.

Dijst, M.J. 2009b. ICT and social networks: towards a situational perspective on the interaction between corporeal and connected presence. In *The Expanding Sphere of Travel Behaviour Research*, ed. R. Kitamura, 45–75. New York: Emerald.

Dijst, M.J. 2013. Space–time integration in a dynamic urbanizing world: current status and future prospects in geography and GIScience. *Annals of the Association of American Geographers* 103: 1058–1061.

Dijst, M. 2014. Social connectedness: a growing challenge for sustainable cities. *Asian Geographer* 31: 175–182.

Dijst, M., and A. Gimmler. 2016. The mobilities of home: towards a new planning for mobilities based on an actor-relational approach. In *Spatial Planning in a Complex Unpredictable World of Change toward a Proactive Co-evolutionary Planning*, ed. G. de Roo and L. Boelens, 150–169. Aldershot: Ashgate.

Duyvendak, J.W. 2011. *The Politics of Home: Belonging and Nostalgia in Western Europe and the United States*. Basingstoke: Palgrave Macmillan.

Ellegård, K., and Svedin, U. 2012. Torsten Hägerstrand's time-geography as the cradle of the activity approach in transport geography. *Journal of Transport Geography* 23: 17–25.

Erskine, R.G. 1998. Therapeutic response to relational needs. *International Journal of Psychotherapy* 3: 234–244.

Erskine, R.G. 2010. *Life Scripts: A Transactional Analysis of Unconscious Relational Patterns*. London: Karnac Books.

Erskine, R.G., and R.L. Trautmann. 1996. Methods of an integrative psychotherapy. *Transactional Analysis Journal* 26: 316–328.

Fischer, J. 2009. Exploring the core identity of philosophical anthropology through the works of Max Scheler, Helmuth Plessner, and Arnold Gehlen. *IRIS: European Journal of Philosophy and Public Debate* 1: 153–170.

Foucault, M. 1986. Of other spaces. *Diacritics* 16: 22–27.

Fox, R. 2006. Animal behaviours, post-human lives: everyday negotiations of the animal–human divide in pet-keeping. *Social & Cultural Geography* 7: 525–537.

Fuchs, T. 2005. Corporealized and disembodied minds: a phenomenological view of the body in melancholia and schizophrenia. *Philosophy, Psychiatry, & Psychology* 12: 95–107.

Giddens, A. 1984. *The Constitution of Society: Outline of the Theory of Structuration*. Berkeley: University of California Press.

Gren, M. 2001. Time-geography matters. In *Timespace: Geographies of Temporality*, ed. J. May and N. Thrift, 208–225. London: Routledge.

Gross, J.J. 1998. The emerging field of emotion regulation: an integrative review. *Review of General Psychology* 2: 271–299.

Guignon, C. 2009. The body, bodily feelings, and existential feelings: a Heideggerian perspective. *Philosophy, Psychiatry, & Psychology* 16: 195–199.

Hägerstrand, T. 1970. What about people in regional science? *Papers of the Regional Science Association* 24: 7–21.

Hägerstrand, T. 1973. The domain of human geography. In *Directions in Geography*, ed. R.J. Chorley, 67–87. London: Methuen.

Hägerstrand, T. 1982. Diorama, path and project. *TESG* 73: 323–339.

Hägerstrand, T. 1985. Time-geography: focus on the corporeality of man, society and environment. In *The Science and Praxis of Complexity*, 193–216. Tokyo: United Nations University.

Hägerstrand, T. 1989. Reflections on "What about people in regional science?", *Papers of the Regional Science Association* 66: 1–6.

Hägerstrand, T. 1995. Action in the physical everyday world. In *Diffusing Geography. Essays for Peter Haggett*, ed. A.D. Cliff, P.R. Gould, A.G. Hoare, and N.J. Thrift, 35–45. Oxford: Blackwell.

Hägerstrand, T. 2009. *Tillvarovävén* (in Swedish). Ed. K. Ellegård and U. Svedin, with a bibliography by B. Lenntorp. Stockholm: Forskningsrådet Formas.

Harrison, P. 2008. Corporeal remains: vulnerability, proximity, and living on after the end of the world. *Environment and Planning A* 40: 423–445.

Harvey, D. 2007. Space as a keyword. In *David Harvey: A Critical Reader*, ed. N. Castree and D. Gregory, 270–286. Malden, MA: Blackwell.

Heidegger, M. 1926/1962. *Being and Time*. New York: Harper & Row.

Heinze, M. 2009. Helmuth Plessner's philosophical anthropology. *Philosophy, Psychiatry, & Psychology* 16: 117–128.

Helm, B.W. 2009 Love, identification and the emotions. *American Philosophical Quarterly* 46: 39–59.

Hogget, P. 1992. A place for experience: a psychoanalytic perspective on boundary, identity, and culture. *Environment and Planning D* 10: 345–356.

Hubbard, P. 2005. The geographies of "going out": emotion and embodiment in the evening economy. In *Emotional Geographies*, ed. J. Davidson, L. Bondi, and M. Smith, 117–134. Aldershot: Ashgate,

James, L. 2005. Achieveness and the meaningfulness of life. *Philosophical Papers* 43: 429–442.

Joye, Y., and Van den Berg, A. 2011. Is love for green in our genes? A critical analysis of evolutionary assumptions in restorative environments research. *Urban Forestry & Urban Greening* 10: 261–268.

Kumashiro, M., C.E. Rusbult, and E.J. Finkel. 2008. Navigating personal and relational concerns: the quest for equilibrium. *Journal of Personality and Social Psychology* 95: 94–110.

Kwan, M.-P. 1998. Space-time and integral measures of individual accessibility: a comparative analysis using a point-based framework. *Geographical Analysis* 30: 191–216.

Lancione, M. 2013. Homeless people and the city of abstract machines: assemblage thinking and the performative approach to homelessness. *Area* 45: 358–364.

Latour, B. 1997. On actor-network theory: a few clarifications plus more than a few complications. *Soziale Welt* 47: 361–381. Available at: www.bruno-latour.fr/sites/default/files/P-67%20ACTOR-NETWORK.pdf (accessed December 29, 2015).

Latour, B. 1999. On recalling ANT. In *Actor Network Theory and After*, ed. J. Law, and J. Hassard, 15–25. Oxford: Blackwell.

Latour, B. 2005. *Reassembling the Social: An Introduction to Actor-Network-Theory.* Oxford: Oxford University Press.

Law, J. 1992. Notes on the theory of the actor-network: ordering, strategy and heterogeneity. *Systems Practice* 5: 379–393.

Law, J. 1993. *Organizing Modernities*, Cambridge, MA: Blackwell.

Lederbogen, F., P. Kirsch, L. Haddad, et al. 2011. City living and urban upbringing affect neural social stress processing in humans. *Nature* 474: 498–501.

Lenntorp, B., 1976. Paths in space-time environments: a time-geographic study of movement possibilities of individuals. Lund: Liber Läromedel/Gleerup.

Lindemann, G. 2009. From experimental interaction to the brain as the epistemic object of neurobiology. *Human Studies* 32: 153–181.

Lindemann, G. 2010. The lived human body from the perspective of the shared world (Mitwelt). *Journal of Speculative Philosophy* 24: 275–291.

Lok, W.-K. 2011. *Foucault, Levinas and the Ethical Embodied Subject.* Amsterdam: Free University (PhD thesis).

Longhurst, R. 2001. *Bodies: Exploring Fluid Boundaries.* London: Routledge.

Mårtensson, S., 1979. *On the Formation of Biographies in Space-Time Environments.* Lund: G.W.K Gleerup.

May, J., and N. Thrift. 2001. *Timespace: Geographies of Temporalities.* London: Routledge.

McQuoid, J., and Dijst, M. 2012. Bringing emotions to time-geography: the case of mobilities of poverty. *Journal of Transport Geography* 23: 26–34.

Merleau-Ponty, M. 1945/2005. *Phenomenology of Perception.* London: Routledge.

Merlo, D.F., R. Filibert, M. Kobernus et al. 2012. Cancer risk and the complexity of the interactions between environmental and host factors: HENVINET interactive diagrams as simple tools for exploring and understanding the scientific evidence. *Environmental Health* 11(Suppl 1): S9.

Miller, H.J. 2005. A measurement theory for time-geography. *Geographical Analysis* 37: 17–45.

Mishare, A.L. 2009. Human bodily ambivalence: precondition for social cognition and its disruption in neuropsychiatric disorders. *Philosophy, Psychiatry, & Psychology* 16: 133–137.

Murdoch, J. 1998. The spaces of actor-network theory. *Geoforum* 29: 357–374.

Orban K., A.K. Edberg and L.K. Erlandsson. 2012. Using a time-geographical diary method in order to facilitate reflections on changes in patterns of daily occupations. *Scandinavian Journal of Occupational Therapy* 19(3): 249–259.

Örmon, K., M. Torstensson-Levander, C. Bahtsevani and C. Sunnqvist. 2015. The life course of women who have experienced abuse: a life chart study in general psychiatric care. *Journal of Psychiatric and Mental Health Nursing* 22: 316–325.

Parker, B. S., and K.S. Dunn. 2011. The continued live experience of the unexpected death of a child. *OMEGA* 63: 221–233.

Perchoux, C., B. Chaix, S. Cummins, and Y. Kestens. 2013. Conceptualization and measurement of environmental exposure in epidemiology: accounting for activity space related to daily mobility. *Health & Place* 2: 86–93.

Perry, B.D. 2002. Childhood experience and the expression of genetic potential: what childhood neglect tells us about nature and nurture. *Brain and Mind* 3: 79–100.

Perry, B.D. 2009. Examining child maltreatment through a neurodevelopmental lens: clinical applications of the neurosequential model of therapeutics. *Journal of Loss and Trauma* 14: 240–255.

Plessner, H. 1928/1975. *Die Stufen des Organischen und der Mensch*. Berlin: De Gruyter.

Puwar, N. 2004. *Space Invaders: Race, Gender and Bodies Out of Place*. Oxford: Berg.

Ratcliffe, M. 2009a. Existential feelings and psychopathology. *Philosophy, Psychiatry, & Psychology* 16: 179–194.

Ratcliffe, M. 2009b. Belonging to the world through the feeling body. *Philosophy, Psychiatry, & Psychology* 16: 205–211.

Rehberg, K.-S, 2009. Philosophical anthropology from the end of World War I to the 1940s and in a current perspective, *IRIS: European Journal of Philosophy and Public Debate* 1: 131–152.

Rice, V.H. 2012. Theories of stress and its relationship to health. In *Handbook of Stress, Coping, and Health: Implications for Nursing Research, Theory, and Practice*, ed. V.H. Rice. London: Sage (second edition).

Richardson, D,B., N.D. Volkow, M.-P. Kwan et al. 2013. Spatial turn in health research. *Science* 339: 1390–1392.

Rose, G. 1993. *Feminism and Geography: The Limits of Geographical Knowledge*. Cambridge: Polity Press.

Schærström, A. 1999. Apparent and actual disease landscapes: some reflections on the geographical definition of health and disease. *Geografiska Annaler* 81B: 235–242.

Schwanen, T. 2007. Matter(s) of interest: artefacts, spacing and timing. *Geografiska Annaler* 89B: 9–22.

Selye, H. 1977. A code for coping with stress. *AORN Journal* 25: 35–42.

Shaw, S.-L. 2012. Guest-editorial introduction: time-geography – its past, present and future. *Journal of Transport Geography* 23: 1–4.

Sheller, M., and Urry, J. 2003. Mobile transformations of "public" and "private" life. *Theory, Culture & Society* 20: 107–125.

Slife, B.D. 2004. Taking practice seriously: toward a relational ontology. *Journal of Theoretical and Philosophical Psychology* 24: 157–178.

Slife, B.D., and F.C. Richardson. 2008. Problematic ontological underpinnings of positive psychology: a strong relational alternative. *Theory & Psychology* 18: 699–723.

Slife, B.D., and B.J. Wiggins. 2009. Taking relationship seriously in psychotherapy: radical relationality. *Journal of Contemporary Psychotherapy* 39: 17–24.

Steppe, K., S. Dzikiti, R. Lemeur and J.R. Milford. 2006. Stomatal oscillations in orange trees under natural climatic conditions. *Annals of Botany* 97: 831–835.

Stewart, L. 2010. Relational needs of the therapist: countertransference, clinical work and supervision. Benefits and disruptions in psychotherapy. *International Journal of Integrative Psychotherapy* 1: 41–50.

Sunnqvist, C., U. Persson, B. Lenntorp and L. Träskman-Bendz. 2007. Time-geography: a model for psychiatric life charting? *Journal of Psychiatric and Mental Health Nursing* 14: 250–257.

Sunnqvist, C., U. Persson, Å. Westrin, L. Träskman-Bendz, and B. Lenntorp. 2013. Grasping the dynamics of suicidal behavior: combining time-geographic life charting and COPE ratings. *Journal of Psychiatric and Mental Health Nursing* 20: 336–344.

Van Os, J., G. Kenis, and B.P.F. Rutten. 2010. The environment and schizophrenia. *Nature* 468: 203–212.

Velleman, J.D. 1999. Love as moral emotion. *Ethics* 109: 338–374.

Wellman, B. 2001. Physical place and cyber place: the rise of personalized networking. *International Journal of Urban and Regional Research* 25: 227–252.

Wild, C.P. 2012. The exposome: from concept to utility. *International Journal of Epidemiology* 41: 24–32.

Wilton, R.D. 1998. The constitution of difference: space and psyche in landscapes of exclusion. *Geoforum* 29: 173–185.

Wirth, L. 1938. Urbanism as a way of life. *American Journal of Sociology* 44: 1–24.

Wittel, A. 2001. Toward a network sociality. *Theory, Culture and Society* 18: 51–76.

Worth, N. 2013. Visual impairment in the city: young people's social strategies for independent mobility. *Urban Studies* 50: 574–586.

7 The time-geographic diary method in studies of everyday life

Eva Magnus

Introduction

This chapter describes and discusses how the time-geographic diary method can engender new knowledge of aspects of everyday life situations that are otherwise difficult to capture, by diarists as well as by researchers and therapists. It is important to learn about the complexity of everyday life in order to reduce constraints and make adaptations to facilitate desirable activities and participation for all individuals.

Everyday life can be defined in various ways. Ellegård (2001) stated that everyday life is shaped by the continuous sequence of activities that the individual engages in 24 hours a day, at any place and with any company. In this way, everyday life is rendered individual and personal, while being influenced both by other people and their choices as well as by the structures and locations of important services in society.

Being engaged in activities, Christiansen and Townsend (2014: 2) have said, is "the occupying of place and time in a rich tapestry of experience, purpose and attached meaning". Activities are parts of the rhythms of everyday life, and they are a primary means by which we organize the world we live in (Hasselkus 2006). What people do has until recently been overlooked as a topic worthy of scholarly attention (Christiansen and Townsend 2014), maybe because daily activities, though often seen, generally go unnoticed (Garfinkel 1964). Frequently, they are presented as ordinary, normal, or basic, such as making food or housecleaning, depicted as of little importance or salience (Hasselkus 2006). Although most people undertake these activities, individuals find their own ways of doing them, whether it is vacuuming the living room, cooking fish for dinner or putting away the laundry. Christiansen and Townsend (2014) have asked how our general, tacit knowledge of what people do can be made more explicit, saying that there is a need to deepen our understanding of how people occupy themselves on an everyday basis. This chapter will propose some answers to this question.

From the perspective of an occupational therapist with a special interest in disability research, this chapter considers how special challenges in everyday life influence the way disabled people live their lives. Even small changes in everyday life activities can have consequences for routines or other activities, consequences that are seen but not noticed. Changes in activities may also affect the balance of everyday life, leading to imbalance and limiting the individual's opportunities for participation in desirable activities (Backman 2010). This chapter presents an overview of how everyday life can be studied by using the time-geographic diary method to seek underlying rules, routines and regulations, going beyond what can be observed and identifying constraints as well as desirable activities in order to increase participation for all in society, regardless of ability.

The time-geographic diary method

The time-geographic diary method (Ellegård and Nordell 1997) is based on and inspired by the time-geographic perspective (Ellegård and Wihlborg 2001; Hägerstrand 1970, 1985), which addresses what people do, where and for how long. The perspective reveals the indivisible individual, and how everyday life is shaped by what individuals prefer and decide to do in the struggle between reaching desirable goals and accommodating various constraints. Although some activities may seem trivial, they can be decisive for performing other types of activities and therefore warrant attention. All the things that people do take time and occupy space, meaning that for their successful performance, both time and space must be considered jointly. When time is of interest, the importance of space is generally neglected. However, the time needed to move from one place to another may be crucial for the individual's decision concerning whether participation in a desirable or necessary activity will be possible.

Ellegård (1994, 1999) started developing the time-geographic diary approach in the 1980s in a study of households' daily life and division of labor, with a focus on what family members were doing at home, elsewhere and in transit. The essence of the method lies in the connections between time, space and activity in a social context in which activities are the center of attention. It emphasizes routines that are shaped by sleep and meals, and preferred and chosen activities in the social and geographic context of the individual. People's time-geographic diaries give a detailed picture of the characteristics of the everyday life of each individual. They are not identical, though there are certain basic resemblances as regards rhythm.

The method relies on a diary organized according to the headings "Time", "What I do (activities)", "Where (places)", "With whom (together with)", and "Comments". The individuals are asked to make notes that are as detailed and complete as possible using their own words. They are to record activities over time in the order in which they occur, starting with waking up in the morning, and including movements, places visited and interactions with other people. Beneath the heading "Comments", people are free to remark on the situations described.

Table 7.1 shows the diary of Agnes (pseudonym), 80 years old, living by herself in a small flat in the city. When writing in her diary, Agnes had just been diagnosed with age-related vision loss, and was waiting for technical aids and other measures to help her live everyday life in a safer, more desirable way. Agnes had few relatives living nearby and her son lived abroad. She was sad about her

Table 7.1 The diary of Agnes, Monday

Time	What I do	Where	With whom	Comments
00:00	Sleeping	At home	By myself	
06:45	Taking a shower	"	"	
07:15	Going shopping	Grocery shop	Shop assistant	
08:00	Breakfast	At home	By myself	
08:30	Tidying up	"	"	
09:00	Listening to the radio	"	"	
10:00	Putting on make-up	"	"	
10:25	Walking to the shop	"	"	
10:45	Doing some errands	Other place for errands	Shop assistant	
11:00	Going for a walk	Outside	By myself	
12:00	Reading the post	At home	"	
12:30	Walking to the shop	Grocery shop	Shop assistant	
13:00	Changing clothes	At home	By myself	
13:10	Washing	"	"	
13:30	Eating dinner	"	"	
14:00	Listening to the radio	"	"	
15:50	Ironing	"	"	
16:55	Preparing a meal	"	"	
18:00	Eating dinner	"	"	
18:15	Nothing special	"	"	
18:50	Listening to the radio	"	"	
20:30	Taking a shower	"	"	
20:50	Going to bed	"	"	
21:00	Listening to the radio	"	"	
22:30	Turning off the radio	"	"	
23:00	Going to sleep	"	"	

difficulties reading and missed perusing international journals and newspapers. Instead, she listened to the radio when at home. Despite vision problems, she managed to go for long walks, even in the winter with snow and ice, as long as she knew her surroundings and did not have to deal with new places. She loved being outdoors and combined walking with visiting shops and places where she could talk to people.

The diaries are individually coded using a code book, in accordance with the diary method (Ellegård 1994; Ellegård and Nordell 1997). The code book classifies the activities (more than 700 activities at five levels of detail) into seven activity spheres (i.e., Care for oneself, Care for others, Household care, Reflection/recreation, Travel, Procure and prepare food, and Work/school), which together constitute the all-embracing project of "Living one's life". All activities, places and persons have number or letter codes that are processed using the Daily Life computer program (Vardagen) (Ellegård and Nordell 1997), making it possible to study daily life from diary notes transformed into graphs and frequency tables.

The computer program presents graphs of the sequence of activities (i.e., everyday activity context), including the individual's movements as well as social and geographic contexts. Frequency tables show how many minutes the individual spends in various activities and places, either alone or with other people. In this way, the frequency table shows the cumulative duration of all occurrences of a given activity, for example, the total minutes spent on all occurrences of the activity "eating" during one day, together with how many times (i.e., the frequency) the individual has engaged in an activity and the maximum, minimum and mean minutes used for each activity. In addition, the graphs show how all activities are distributed throughout the day as well as the duration of each occurrence of an activity, forming the daily routines of the individual (in contrast to the added time use, this is the real time use).

Figure 7.1 presents three graphs from Agnes' diary; her activity oriented individual path shows occurrences of all her activities in sequence, including travel activities. The broad vertical parts of the path show her activities, while the thin horizontal lines just indicate change of activity. This is her everyday activity

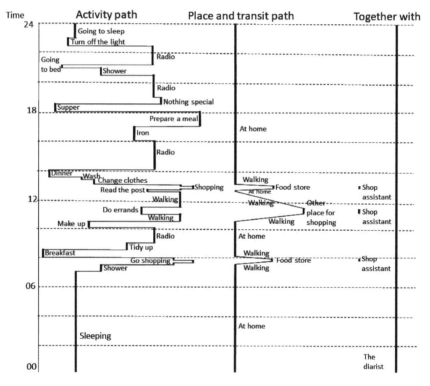

Figure 7.1 The graphs of the diary of Agnes. Her activity oriented individual path (left) illustrates each activity she performed in sequence. Her changing from one activity to another is indicated by thin horizontal lines. The second graph (middle) shows places she visited and the third graph (right) shows the persons she spent time with that day.

context, to the left. The middle part of the figure shows places where she spent time and time used for transportation, the geographic context (which corresponds to the conventional individual path). The social context, to the right indicates her being by herself and in company with other people. The graphs illustrate the indivisible individual, in which she can recognize herself and her everyday life. Time is shown on the vertical axis, while the codes assigned to activities, places and persons are on the horizontal axis.

Agnes' individual activity path shows that she spent time listening to the radio and going out for walks and errands, interrupted by meals and self-care activities. The graph of her geographic context shows that she went out of the house three times, walking to shops for errands. The graph of the social context shows that when she went out, she noted spending time with shop assistants; the rest of the day she spent by herself.

Studying Agnes' diary notes for one whole week gives an image of her everyday life with its routine activities, such as taking a shower, often twice a day, putting on make-up, preparing food and eating, walking to the shop nearby or to the city center, and listening to the radio. Frequency tables make it possible to calculate her time use for selected activities, and to compare her time use for different activities and in social and geographic contexts with that of others of her age.

Studies using time-geographic diaries

Understanding students and the physical activities of older men

The first major study using the time-geographic diary method in a therapeutic context concerned Scandinavian students of occupational therapy and physiotherapy. The time-geographic diary method was used similarly to how it was in the case of Agnes, cited above, revealing the repertoire of everyday activities of occupational therapy and physiotherapy students in Scandinavia (Alsaker et al. 2006). The students were asked to keep diaries for one week. In total, 154 diary days were studied, and the results captured the students' time use for activities, in places, and with other people or by themselves during that week. It was evident that the students organized their daily life with a focus on performing activities in three of the seven main categories: Studies, Care for oneself, and Reflection/recreation. Care for oneself includes physical exercise, sleeping, and eating. The activity pattern changed on the weekend, less time being spent on Studies and more time on Reflection/recreation and Care for oneself, mainly more sleep. Reflection/recreation was characterized by spending time with other students. Although the students spent time with fellow students and friends on the weekend, this was also the case during the week. Unsurprisingly, the study illustrated how everyday life within a group of students differed in time used for studies as well as in time spent with others. Some students spent most of their free time by themselves while the majority socialized in most of their free time. What the students thought about the activities they chose to do, how they socialized, and where they went remains unknown. Studying the frequency table revealed that the students engaged in surprisingly few types of activities, which were repetitive and closely linked to the overall pattern of the culture.

The diary method has also been used to capture elderly men's physical activity patterns in everyday routines (Bredland et al. 2015). Of special interest was time use in various types of physical activities, such as the degree to which the participants did recommended exercises and whether they were physically active as part of their daily routines. In this study, the diary method was combined with the use of metabolic equivalent intensity levels (METS) in the analyses. METS intensity codes (Ainsworth et al. 2000) make it possible to compare different types of physical activities. A METS score of more than 3 is classified as moderate intensity, exemplified by light- or moderate-effort weight lifting. Chopping wood and shovelling snow by hand are classified as 6 METS. The elderly men were asked to keep a diary for one week, after which the activities with METS scores of more than 3 were combined with exercise data to determine their physical activity patterns. By combining the diary method with METS data, this study made it possible to illustrate how physical activity could be part of ordinary everyday life activities, and not necessarily just from formal physical exercise. The elderly men achieved the recommended amount of physical activity in daily activities at home, in the garden, and when playing with grandchildren as well as when exercising at a fitness center. The interesting conclusion of this study is that the diary method allowed physical activity to be studied in an extensive way, taking account of how physical activity is woven into and made part of ordinary daily activities.

In these two studies, i.e., those of Alsaker et al. (2006) and Bredland et al. (2015), the diary method with graphs and frequency tables was used to answer overarching questions about the activity patterns and time use of the selected groups. The method gave rise to new knowledge as well as new questions, indicating other opportunities to apply the diary method. The time-geographic perspective and the application of the diary method highlight what appears physically and materially. When looking at an individual's diary, questions of a subjective nature often arise, such as: Was this an ordinary day or week? Do the activities recorded in the diary correspond to the activities that this person likes to do or has to do? What activities were omitted and why? Were the persons the individual met during the day also those whom s/he preferred to spend time with, and in these places? Did the person usually travel to school or work in the way recorded in the diary?

The activity pattern tells us about the regularities and content of daily life, but not about priorities, motives, barriers or the degree to which individuals are unable to make optimal choices. To address this deficiency, the diary method is often complemented with other qualitative methods, such as interviews, focus groups and observations, to gain further information and a better understanding of what is and is not recorded in the diary. This combined approach will be considered in the next section.

The everyday life of disabled students revealed in time-geographic diaries and interviews

The study of the activity patterns of Scandinavian occupational therapy and physiotherapy students (Alsaker et al. 2006) was a direct inspiration of Magnus' (2009) study of the everyday life of disabled students in higher education. This Scandinavian study raised questions about the activity patterns of a group of disabled students, about how they lived their everyday lives, combining the time-geographic diary method with in-depth interviews and focus groups. Disability was understood in line with the Nordic relational model (Gustavsson et al. 2005), emphasizing the need to adapt the environment to accommodate a broader range of people.

Higher education is one of the most important strategies to increase the participation of disabled persons in society (Molden et al. 2009; Vedeler 2009). Research demonstrates that disabled students encounter hindrances in all aspects of their study situations on campus (Brandt 2010; Fuller et al. 2004; Magnus and Tøssebro 2014; Stamer and Nielsen 2008) and in fieldwork, for example, in buildings, transport, equipment and attitudes (Fuglerud and Solheim 2008; Jung et al. 2008; Nolan et al. 2015). This is problematic because it counteracts their participation on equal terms, which is a societal goal.

Over the last 30 years, equality and participation have been the main political goals for disabled people in Norway, as in other countries. There has been a general attitudinal shift concerning disability in higher education as in other parts of society, as more attention has been directed towards environmental barriers and universal design. In 2003, a Norwegian government resolution stated that the proportion of disabled students in higher education should increase to equal that of the rest of the population (White Paper 40 2002–2003). The Norwegian Higher Education Act (2005) defines students' rights to appropriate accessibility, and universities are instructed to take the necessary steps to ensure that disabled students have the same access to education as their non-disabled peers. Also, the Norwegian Anti-discrimination and Accessibility Act (2008) emphasizes the responsibility of universities to ensure equal educational opportunities for disabled persons.

Being a student often entails living by oneself for the first time, often after moving to another part of one's country. For most students, accommodating new educational expectations in addition to handling food preparation, housework and budgeting for the first time can be challenging. In line with this, the intention of Magnus (2009) was to improve our knowledge of the connections between disability and active participation in student life. Student life involves influences from both inside and outside the university, which combine to constitute everyday life. Central questions addressed by Magnus (2009) were: How do students with disabilities spend their time? To what degree do their daily lives differ from those of non-disabled students? What activities do they omit, and what are the consequences for the individual? What kind of constraints do these students face?

How do these limit desirable participation, and how do the individual disabled students handle this? What conditions promote participation, and what meaning does participation have?

Thomas (1999) claimed that instead of studying the individual one should look into the aspects of society that help shape the experiences of disabled persons. However, the individual perspective can be vitally important in exposing the constraints confronting the disabled, together with their implications. Some constraints are obvious, while others are discerned only by those who know them from their own experience. That was why it was important to explore, in depth, the everyday life experiences of the students with disabilities (Magnus 2009).

Nineteen students, aged 20–44 years, participated in the study. They studied different subjects at one university and one university college in a Norwegian city. Their impairments included limited mobility, hearing impairment, visual impairment, hidden and chronic diseases, and dyslexia. Each student kept a diary for one week.

The diary-keeping was followed by individual in-depth interviews (Gubrium and Holstein 1997; McCracken 1988), which starting by addressing subjects recorded in the diaries, followed by themes from a predetermined interview guide. The diaries functioned to steer the exchanges (McCracken 1988), stimulating further descriptions, comments, reflections and associations regarding daily life as recorded in the diary.

At the end of the research process, common topics identified in the interviews were discussed in focus groups (Krueger and Casey 2000), three of which met three times each. This process was based on the idea of gradually building on the overview of everyday life in the diaries, deepening the understanding of the participants' experiences in the next two steps. The individual interviews and the focus group discussions were audio-recorded and transcribed verbatim.

The diaries illustrated how the everyday lives of these disabled students both differed from and resembled those of the non-disabled occupational therapy and physiotherapy students in the Scandinavian study by Alsaker et al. (2006). Their complex everyday lives were captured in their diaries, and although they had impairments that differed in their extents and consequences, certain experiences were shared.

Projects

Everyday life consists of many activities that are parts of various *projects*. In time-geography, a project consists of all the specific activities a person undertakes to achieve a goal (Ellegård and Wihlborg 2001). Everyone's overarching project is living one's life, and within that context people are involved in all kinds of long- and short-term projects, such as preparing food, getting an education, caring for children and gardening. These projects differ in their durations and the places involved. They can be of short duration, such as inviting friends for Saturday evening, or of long duration, such as obtaining a professional education. One and the same activity can be part of many projects; for example, playing handball can be part of the projects "keeping fit" and "being with friends".

The study results indicated that the students' everyday lives consisted of three main projects (Magnus 2009):

1) qualifying for employment,
2) accomplishing everyday life tasks, and
3) taking care of health.

Qualifying for employment. Although most of these students were studying part time, they spent the same number of hours studying per week as did average full-time Norwegian students, i.e., 30 hours. Some of them spent more than the average time on their studies. Most of the disabled students needed extra time for reading, writing and doing practical things between educational activities that other non-disabled students could do more quickly, such as finding books in the library, getting accessible literature, using copy machines, taking notes or moving between buildings and rooms. Students with low vision noted that it is more time-consuming to use an audiobook than an ordinary book, because it is difficult to skim.

Taking care of health involved activities such as physiotherapy and other treatments, but also activities connected to well-being, pleasure, and relaxation, such as reading books, seeing films with friends, handicrafts, physical activities, and baking. One clear finding is that the disabled students spent more time resting and sleeping than did the occupational therapy and physiotherapy students in general (Alsaker et al. 2006) or what was average in Norway. This greater need for rest and sleep can be seen as resulting from impairments that made educational and ordinary daily activities more time-consuming and tiring than they were for other students. Other reasons concerned diagnoses that caused pain or fatigue due to medication. Taking care of health was crucial so that these students would be able to study and reach their goals.

Accomplishing everyday life tasks. Educational activities involved various challenges and extra time use, and this was also the case when it came to ordinary daily activities that all people do. This was time-consuming for some students, especially those who were dependent on assistance offered by service organizations for doing everyday life activities for themselves at home. They had to be sure that their environments were properly adapted to allow them to study.

Heidi, for example, used to make notes on all the everyday life activities that had to be done. She had severe movement restrictions and used personal assistants at home and at university. These notes made all the important activities clear to both the assistants and herself, and gave her a feeling of control. She had to have structured plans and everything in place to ensure that she could manage.

> All the special equipment I need, the urine bag and everything, has to be in place. I have lists of everything, so that I know that I have it all. And if there is a mess I get sweaty, because I feel that I am not good enough.

For some students, accomplishing everyday life tasks was a never-ending job. Sandra was divorced and had a daughter attending a day nursery. She had post-polio syndrome and experienced difficulties walking; she managed the challenges at home by prioritizing what was important to do.

> There are many things I can manage during the day, such as preparing food and tidying up after dinner. But things like cleaning the flat have to wait, and these things give me an unpleasant feeling. I have to prioritize my daughter, be sure that she and her clothes are OK, and to take care of myself. Vacuuming – that is a big job and hard work, as it is to change bedclothes, which is a nightmare. At the same time, it is so pleasant to go to bed in clean bedclothes.

The extra time needed for many activities during the day had consequences for what could be done during the rest of the day.

Consequences. The students in this study had everyday lives that both resembled and differed from those of the other students. Disabled students spent more time resting and sleeping, on ordinary daily activities, alone, and planning their days, and they spent at least as much time studying as did the other students, even though most of them studied part time. Consequently, they developed strategies for addressing these challenges, by:

1) prioritizing activities in proportion to their energy level on a given day or to what they regarded as realistic;
2) planning structure and routines to create predictability; and
3) slowing study progression to what was achievable.

The transition to higher education can make many everyday life activities more palpable to most students because they must manage them by themselves for the first time. They have to focus on study activities, domestic work, and making new friends. For disabled students, it is crucial that the learning environment be accessible. To participate in learning and socializing, accessible housing, transport, welfare services, educational financing, and practical support must all be available as needed.

Constraints

When undertaking activities in daily life the individual will experience *constraints* that limit what is possible for that person to do. Three kinds of constraints are described in time-geography (Hägerstrand 1985):

*   capability constraints – the limits imposed by the individual's biological, economic, material, cognitive and mental resources;
*   coupling constraints – the individual's need for tools, aides and assistance at given times and places; and
*   authority constraints – rules and laws to be followed, set by an authority.

The disabled students experienced constraints of all these kinds. Descriptions of their projects illustrated the consequences of impairments, their capability constraints leading to coupling constraints despite the rules formulated to facilitate life for these students.

These students described learning as a pleasure and said that their long-term aim was employment and independent living. The other side of their motivation was a desire to show other people that they could manage, despite explicit or implicit negative expectations. One male student, Håkon, said:

> I want to show society that I can contribute despite the problems I have, and that I am no different from others. So if people think that they can break me down, they are wrong.

Håkon had to demonstrate that he had the will and competence to participate despite impairments that others thought would stop him. He was happy about his studies and about being a student. At the same time, he had worked hard to get the adaptations he needed to reduce his capability constraints, and to find a way to balance studies with daily activities, friendship, rest and physiotherapy treatment. He had problems getting the funding he needed (authority constraints) and found it difficult to carry out his household chores in the flat he shared with three other students. He considered hiring someone to do his part of his chores.

Although there are political agreements in Norway on making it possible for more disabled individuals to study, these students with capability constraints met other constraints making their everyday lives more challenging than those of other students. There were visible barriers on campus, such as stairs and heavy doors, and other less obvious barriers experienced only by persons with disability, such as long distances between buildings, a lack of signage about accessibility in lifts and on bulletin boards, inaccessible websites, and lecture rooms with no or insufficient space for wheelchairs or other necessary equipment. These are all outcomes of constraints set by authorities. These constraints made moving around and finding information difficult and more time-consuming. Adaptive literature, such as audiobooks, is reserved for blind and extensively visually impaired students. However, students with less extensive visual impairments or dyslexia also find reading much easier with audiobooks. Due to legislation (authority constraints), however, they did not have the same rights to these resources: they were permitted to borrow them, but only literature that was already produced.

Universal design of learning environments involves both teaching and assessment. When lecturers display a negative attitude or refuse to use the equipment one student needs in order to participate, new constraints arise. For example, one lecturer gave a student the impression that s/he disliked the student's use of a tape recorder during the lecture. Other lecturers would give lecture notes to the student with cerebral palsy but not to the student with dyslexia, demonstrating that some problems were deemed real and to be taken seriously while others were not. One could say that these problems were about attitudes, but I will call them coupling constraints, making it difficult for people to meet and understand each other.

Kari had hearing aids in both ears and brought a frequency modulation device to her lectures, which meant that her professors had to use a microphone connected to her hearing aids. Although she asked the teachers to use the microphone so that she could hear, they did not always do so, even though she explained its importance for her learning. For Kari, assistive technology was crucial so she could participate in the same way as other students. The technology that should have increased her ability had no such effect when the people she dealt with refused to use it. In this way, they created coupling constraints for her, maintaining her capability constraints.

Restrictions encountered outside higher education institutions also influence the kind of study situation the individual will have. Students who need home care services and special transport are dependent on the care services' ability to be on time and are therefore confronted with coupling restrictions. In the diary of one female student, "waiting for a taxi" was noted several times during the day. She explained that she had to plan for the taxi coming late, but also had to be ready if the taxi happened to come on time. Another example of coupling restrictions is the home helper who has to show up when the student is at home. If not, the student risks not getting the help needed. Personal assistance is crucial for those having access to it, but the students spoke of the limited hours allowed (authority constraints) versus what they needed (capability constraints).

In addition, financial support is an authority constraint for many disabled students. Educational funding is of particular importance for students at risk of taking longer than is standard to complete their studies, especially in a situation where work for pay is hard to handle. Funding is not adapted to the life situation of disabled students, even though they need more time for their studies than do other students.

For all students, friendships and developing a social network are of special importance. The disabled students talked about limited opportunities to participate in social life. There were problems getting to the venues where events were happening, problems entering and exiting after arriving there, and problems getting around inside. Students needing time to plan ahead faced real challenges because socializing usually happens spontaneously (coupling restrictions). Socializing was therefore often not prioritized by students with disabilities because it demanded too much from them, in time and energy.

In this study, the time-geographic diaries were used to obtain an overview of the everyday lives of the disabled students. They were an entry point for learning about their activities and projects in particular social and geographic contexts, creating a basis for discussions and further analyses. The diaries were used in combination with in-depth interview methods to get detailed information about what had been noted, about meaning, constraints and opportunities. In addition, the students themselves reflected on how everyday life challenges influenced what activities were regarded as desirable, because they had to consider what their consequences might be.

Employed women with rheumatoid arthritis

Similar combinations of the time-geographic diary method with other methods have been used in studies of the everyday life of other groups of participants. Living with chronic disease often involves constraints on everyday life activities because of pain and tiredness. Jakobsen (2004) used the time-geographic diary in combination with focus groups to build knowledge of the everyday lives of employed women with rheumatoid arthritis. For one week, 11 women kept diaries, some recording just their main activities during the week, others describing their doings in greater detail. Based on analyses of the diaries, various subjects were discussed in focus groups, such as time use for different activities during the days and week, expectations the women encountered as employees, the meanings, opportunities and duties connected to work, opportunities to "scale down" in accordance with one's body, accommodations made in daily life, and initiatives to make employment easier for persons with chronic diseases. The women described how capability constraints due to the disease meant various challenges at work and home. Bodily limitations made full-time work difficult. In addition, they were reluctant to talk about pain and tiredness at work, because they were afraid of being regarded as a burden on the working environment. Some of them reflected on reducing their working hours to take better care of their health, but this raised the specter of authority constraints, as disability insurance required at least 50% reduced working capability, and that was more than they needed. In this way, the authority constraint made it difficult for these women to strike a balance between work, training and treatment, while retaining sufficient energy to be social with family and friends. In addition, the individual would have been helped to accommodate both work and health requirements if there were fewer restrictions and more flexibility at the workplace.

The methods used in this study developed detailed knowledge of the everyday lives of a group of women active in the workforce but not often recognized because their impairments are invisible. Of special interest in this and similar studies that use the diary method is the focus on the full 24-hour day, not only the hours the persons are working or studying. As in the study of disabled students, the whole day must be examined to get a comprehensive picture of the coupling constraints that severely complicate everyday life for individuals with impairments. Circumstances at work, on campus or at home influence the rest of the day.

The everyday life of breast cancer survivors

In an ongoing study of the everyday lives of breast cancer survivors seven years after treatment, time-geographic diaries and in-depth interviews were combined in order to describe the health-promoting everyday life projects of these women (Magnus et al. forthcoming). In this study, the interviews were read and their contents were highlighted as they described central and desirable activities and constraints in everyday life. Highlighted passages from each interview were then compiled in a short synthesized text summarizing each interview. By analysing this text, groups of activities could be identified.

Projects were identified by combining the activities described in both diaries and interviews with the meanings of the activities expressed in the interviews. In the end, four central everyday life projects were generated:

1)   doing what is good for my body and soul;
2)   creating something that is nice and joyful;
3)   keeping my family and social network together; and
4)   keeping my mind active.

Spending time on these projects also involved coping with constraints. Many of these women had been heavily influenced by previous cancer treatment (capability constraints), and struggled with sleeping, concentrating, remembering, pain, fatigue, etc. This meant that they had to prioritize how they spent their time, to be able to undertake desirable activities and activities with special meaning, together with the people they wanted to meet, and to get necessary rest. Their everyday life projects had to balance activities and rest, and in this way promote health.

As seen in the studies described (i.e., Jakobsen 2004; Magnus 2009; Magnus et al. forthcoming), a combination of the time-geographic diary method with other qualitative methods gives new opportunities for a deeper understanding of the everyday life situation of the participants.

Combinations of interviews and time-geographic diaries

Løvmo (2012) reflected on the differences in how the interviews in her study took place. The purpose of her study was to build knowledge of the everyday lives of disabled mothers. Eight women participated; four of them kept diaries for one week, but all participants were interviewed. The results indicated that the everyday lives of these women were largely similar to those of other women. They also illustrated how these women faced challenges connected to the impairment itself (capability constraints) and to environmental restrictions. Løvmo (2012) reflected on how the women who had kept the diaries were better prepared for the interviews: they had spent time thinking about their situations through making notes about them, and consequently had more to say. In addition, after having studied the diaries, Løvmo herself had more knowledge of the everyday lives of these women, and had more questions about the situations described in them.

Consciousness-raising

It is evident when using the diary method that keeping a diary for several days raises the diarist's consciousness of various aspects of his/her everyday life. One female disabled student (Magnus 2009) discovered how much time she spent organizing her everyday life. Suddenly she understood why she had to put aside building a social network: there was simply no time or energy left for it, and not enough hours of assistance to manage it. She wanted to bring her time-geographic diary graphs to the Norwegian Labour and Welfare Organization (NAV) in hopes of making them understand her challenges and her need for more hours with personal assistants. In her case, the frequency table was used to calculate additional time spent administering the assistants over six days.

This female student was not the only one who gained a better understanding of how everyday life was shaped by activities in time and space. One male student wrote at the end of his week of keeping a diary that he realized that he was actually not lonely (Magnus 2009). He had many good friends that he spent time with, when he had time and was not too tired. His reflections were based on the observation that his notes during that week could give the impression that he lacked friends. During the interview, he talked about the activities he liked to do with friends, and about how walking long distances with them could be a challenge. Another student saw how time spent sleeping had consequences for other parts of the 24-hour day. He had cerebral palsy and got extremely tired from his studies and various necessary daily activities.

In these cases, the diary method illustrated the importance of studying the whole day and its sequence of activities in order to understand the conditions affecting people's everyday lives. Although the students had already prioritized how they spent their time, the diary helped them visualize the connections between time use, constraints met and the 24 hours of the day. These reflections were later reiterated by organizations for disabled persons, which commented that Magnus' (2009) findings were important and confirmed what they already knew. For these organizations, the time-geographic diary method was a way to raise consciousness and understanding among politicians and other authorities of the challenges of living an everyday life with impairment. This case illustrates how the time-geographic diary method can provide a way to describe people's everyday lives in a new and reliable way and, it is hoped, initiate change so that authority constraints are eased.

Research into the everyday lives of various groups of people may involve questions about challenges in daily life. All individuals confront limitations in life, and the time-geographic perspective together with the time-geographic diary method offer concepts for elucidating the various kinds of these limitations. When limitations are discovered, the next step is for the individuals themselves or the authorities to make decisions on how to address them.

Reflections on time use are also found in the master's thesis of Solli (2017), who explored how participants and their supervisors in a program of lifestyle changes made use of time-geographic diaries to start a change process in everyday life. Eleven participants kept diaries for seven days. They were surprised by their own activity patterns, at how much time they spent in front of the television or PC, involved in sedentary activities. They said that they thought they walked more during the day than they actually did. Some were surprised that they did less than expected, while others were surprised by the number and density of their activities – there was no rest. This study illustrates how the time-geographic diary made the participants more aware of their everyday life activities, and of how these could be connected to capability constraints. Time spent in front of the television and with social media also had consequences for the participants' social lives, which became clear when they looked into who they spent time with.

The diary method in adjusting the everyday life of an individual

The time-geographic diary method can be used when discussing with a person what s/he wants everyday life to be like, for instance, when s/he is involved in a program of lifestyle change or coping with chronic illness and having to accommodate capability constraints. When examining and talking about the person's diary, the following may be relevant questions: What would you like your day to be like? What activities are the most important for you? Who do you want to spend time with, and what places are important for you to be able to reach or stay at? What things make you the most tired, and when does this happen? What constraints are the most important in prompting you to make changes?

In Western countries, stress and mental health problems due to excess activities (or the opposite) have consequences for the balance of everyday life. In the early 1900s, Adolf Meyer (Franklin 1922), the founder of occupational therapy, stated the importance of a balance between work, sleep and rest for health. To reach what the individual regards as a desirable balance between activities in a given social and geographic context, the individual has to be conscious of his or her own everyday life. This is a central issue when working on lifestyle changes, health promotion and rehabilitation. Solli (2017) found that using the diary as a starting point when discussing lifestyle changes with persons facing health challenges made it easier to suggest actions that could be meaningful for the individual. In addition, when using the diary, actions turned out to be more realistic in terms of what the participants could actually handle. The participants' supervisors reflected on how they had earlier asked the participants to do more than they now understood would be possible for them. Now they had a more realistic grasp of the participants' everyday lives and could discuss actions that were more individualized, making it possible for the individual to succeed. The supervisors also reflected on how the diary method made it easier for them to ask questions they previously did not have the courage to ask. The diary legitimized asking about minor physical activities and time spent in front of a screen, because the participant had written about them. As a consequence of knowing more about the everyday lives of the participants, the supervisors placed more emphasis on everyday life activities and less on physical activities when planning actions.

When working on changes in people's lives, the diary method highlights new elements and new possible strategies for health professionals to consider. The diary visualizes time use and important activities, making it possible to discuss constraints, other critical elements, and what constitutes a desirable everyday life.

Health professionals often discuss capability constraints with their clients, but other constraints can be as important to consider. As demonstrated by the studies of disabled students, the coupling constraints limit their opportunities to undertake desired activities. Lifestyle changes usually involve other people in a household or workplace, as is the case when working with rehabilitation. Irrespective of whether changes concern lifestyle or a rehabilitation process, the actions have to do with everyday life activities in one way or another. Changes in everyday life involve the social networks and physical environment of the individual. In one way or another, these influence family members and others, which means that coupling constraints can be a challenge to be addressed. It is also necessary to consider knowledge of what may represent authority constraints. The time-geographic diary can make invisible restrictions visible for the individual affected, helping professionals, as well as for leaders in positions of authority.

The time-geographic diary method has so far been a time-consuming method to use, due to the manual coding and analysis of what the individual has written. New technology is coming whereby the individual can make notes directly into the program installed as an app in a cell phone. This will make the diary even more useful, as the individual will retain power over the material.

It can be claimed that the individual does not record the whole truth in a diary, that a lot of information about everyday life is left out. That is right, but that is also the case with other methods. What is central to the time-geographic diary method is that the individual decides what to record, and those are often the things that are important to him-/herself – it is not the supervisor, researcher or therapist who decides what are important elements in the life of another person. In this way, the time-geographic method can be used to emphasize central ideas about empowerment and support a bottom-up therapeutic strategy.

Conclusion

This chapter has paid attention to how the time-geographic diary method has been used in research and health promotion work in Norway. The method opens up new opportunities when seeking new knowledge of how people live their everyday lives with or without impairments, and of how to visualize daily life activities that otherwise are difficult to grasp, such as constraints that have consequences for desirable participation in society. Although the method is new for health professionals, results indicate that it has potential in health areas that concern everyday life and people's opportunities for participation, whether the aim is lifestyle changes or an individual's adaptation to a new situation due to illness or impairments. With its emphasis on the individual's own descriptions of everyday life, the diary method is a unique way of raising consciousness of the person's own life, the constraints s/he faces, and what is considered a desirable everyday life. This can be seen as an important starting point for introducing changes or making adaptations in order to ease constraints, thereby contributing to achievement of the goal: participation for all.

References

Ainsworth, B.E., Haskell, W.L., Whitt, M.C., et al. 2000. Compendium of Physical Activities: An Update of Activity Codes and MET Intensities. *Medicine and Science in Sports and Exercise*, 32(9): 498–516.

Alsaker, S., Jakobsen, K., Magnus, E., et al. 2006. Everyday Occupations of Occupational Therapy and Physiotherapy Students in Scandinavia. *Journal of Occupational Science*, 13(1): 17–26.

Backman, C.L. 2010. Occupational Balance and Well-Being. In Christiansen, C.H. and Townsend, E.A. (eds), *Introduction to Occupation: The Art and Science of Living*. Upper Saddle River, NJ: Pearson Education, pp. 231–249.

Brandt, S.S. 2010. *Tilretteleggingsutfordringer i høyere utdanninger før og nå: en studie av tilrettelegging i høyere utdanning i lys av lovendringer og målsetting om å sikre høyere utdanning for en mer mangfoldig studentmasse* [Challenges in Adaptation in Higher Education Before and Now: A Study of Adaptation in Higher Education in the Light of Legislation Changes and Goals to Ensure Higher Education for Diversity of Student Population]. Oslo: NIFU STEP.

Bredland, E.L., Magnus, E., and Vik, K. 2015. Physical Activity Patterns in Older Men. *Physical & Occupational Therapy in Geriatrics*, 33(1): 87–102.

Christiansen, C.H., and Townsend, E.A. 2014. An Introduction to Occupation. In Christiansen, C. & Townsend, E., *Introduction to Occupation: The Art of Science and Living* (2nd edn). Harlow: Pearson Education, pp. 1–34.

Ellegård, K. 1994. *Att fånga det förgängliga: Utveckling av en metod för studier av vardagslivets skeenden. Vardagslivets komposition delrapport 2*. Occasional Papers 1994:1. Kulturgeografiska institutionen, Göteborgs universitet.

Ellegård, K. 1999. A Time-Geographical Approach to the Study of Everyday Life of Individuals: A Challenge of Complexity. *GeoJournal*, 48(3): 167–175.

Ellegård, K. 2001. Att hitta system i den välkända vardagen: En tankeram för studier av vardagens aktivitetsmönster och projekt. In Ellegård, K. and Wihlborg, E. (eds), *Fånga vardagen: Ett tvärvetenskapligt perspektiv*. Lund: Studentlitteratur, pp. 41–66.

Ellegård, K., and Nordell, K. 1997. *Att byta vanmakt mot egenmakt* [To Change Powerlessness by Empowerment]. *Metodbok*. Stockholm: Johansson & Skyttmo förlag.

Ellegård, K., and Wihlborg, E. 2001. Metoder för att studera och analysera vardagen. In Ellegård, K. and Wihlborg, E. (eds), *Fånga vardagen: Ett tvärvetenskapligt perspektiv*. Lund: Studentlitteratur, pp. 5–12.

Franklin, M.E. 1922. The Philosophy of Occupational Therapy. (Arch. of Occupat. Ther., February, 1922.) Meyer, Adolf. *British Journal of Psychiatry*, 68(283): 421–423. DOI: 10.1192/bjp.68.283.421.

Fuglerud, K.S., and Solheim, I. 2008. *Synshemmedes IKT-barrierer: Resultater fra undersøkelse om IKT-bruk blant synshemmede* [The ICT Barriers for Partially Sighted and Blind Persons: Results from a Study of the Use of ICT among Partially Sighted and Blind Persons]. Oslo: Norsk Regnesentral.

Fuller, M., Healey, M., Bradley, A., and Hall, T. 2004. Barriers to Learning: A Systematic Study of the Experience of Disabled Students in One University. *Studies in Higher Education*, 29: 303–318.

Garfinkel, H. 1964. Studies in the Routine Grounds of Everyday Activities. *Social Problems*, 11(3): 225–250.

Gubrium, J.F., and Holstein, J.A. 1997. *The New Language of Qualitative Method*. Oxford: Oxford University Press.

Gustavsson, A., Tøssebro, J., and Traustadóttir, R. 2005. Introduction: Approaches and Perspectives in Nordic Disability Research. In Gustavsson, A., Sandvin, J.R., Traustadóttir, R. & Tøssebro, J. (eds), *Resistance, Reflection and Change: Nordic Disability Research*. Lund: Studentlitteratur, pp. 23–44.

Hägerstrand, T. 1970. What About People in Regional Science? *Regional Science Association Papers*, 24: 7–21.

Hägerstrand, T. 1985. Time-Geography: Focus on the Corporeality of Man, Society, and Environment. In *The Science and Praxis of Complexity*. Tokyo: United Nations University, pp. 193–216.

Hasselkus, B.R. 2006. The World of Everyday Occupation: Real People, Real Lives. *American Journal of Occupational Therapy*, 60: 627–640.

Jakobsen, K. 2004. *Hvordan arbeidsdeltakelse etter evne kan muliggjøres for kvinner med revmatisme*. Sluttrapport fra prosjektet Yrkesaktive kvinners dagligliv. Rapport nr.1, Trondheim: Høgskolen i Sør-Trøndelag, Avdeling for helse- og sosialfag. Master's thesis. Trondheim: Norwegian University of Science and Technology.

Jung, B., Salvatori, P., Tremblay, M., Baptiste, S., and Sinclair, K. 2008. Inclusive Occupational Therapy Education: An International Perspective. *WFOT Bulletin*, 57: 33–41.

Krueger, R.A., and Casey, M.A. 2000. *Focus Groups: A Practical Guide for Applied Research* (3rd edn). London: SAGE.

Lov om forbud mot diskriminering på grunn av nedsatt funksjonsevne. 2008. The Act on Accessibility and Discrimination. Oslo: Barne- og likestillingsdepartementet.

Lov om universiteter og høyskoler. 2005. The Higher Education Act. Oslo: Kunnskapsdepartementet.

Løvmo, R. 2012. *Mødre med nedsatt funksjonsevne og deres hverdag* [Disabled Mothers and Their Everyday Life]. Master's thesis. Trondheim: Norwegian University of Science and Technology.

Magnus, E. 2009. *Student, som alle andre: En studie av hverdagslivet til studenter med nedsatt funksjonsevne* [Student, like Everybody Else: A Study of the Everyday Life of Students with Impairments]. PhD thesis. Trondheim: Norwegian University of Science and Technology.

Magnus, E., and Tøssebro, J. 2014. Negotiating Individual Accommodation in Higher Education. *Scandinavian Journal of Disability Research*, 16(4): 316–332. Available at: http://dx.doi.org/10.1080/15017419.2012.761156 (accessed June 2018).

Magnus, E., Jakobsen, K., and Reidunsdatter, R. (forthcoming). Everyday Life Projects in Breast Cancer Survivors Seven Years after Treatment.

McCracken, G. 1988. *The Long Interview*. London: SAGE.

Molden, T.H., Wendelborg, C., and Tøssebro, J. 2009. *Levekår blant personer med nedsatt funksjonsevne: Analyse av levekårsundersøkelsen blant personer med nedsatt funksjonsevne 2007* [Living Conditions by Persons with Impairments: Analyses of the Study of Living Conditions among Persons with Impairments 2007]. Trondheim: NTNU Samfunnsforskning AS.

Nolan, C., Gleeson, C., Treanor, D., and Madigan, S. 2015. Higher Education Students Registered with Disability Services and Practice Educators: Issues and Concerns for Professional Placements. *International Journal of Inclusive Education*. 19(5): 487–502.

Solli, M.S. 2017. *Tidsgeografisk dagbok i Frisklivsresept* [Time-Geographic Diary in Health Promotion Work). Master's thesis. Trondheim: Norwegian University of Science and Technology.

Stamer, N.B., and Nielsen, S.B. 2008. *"Vi er jo ikke en del af universitetets bevidsthed . . ."*: *En undersøgelse af barrierer for studerende med handicap på de lange videregående uddannelser* ["We are Not a Part of the Consciousness of the University . . .": A Study of Barriers for Students with Disability at Universities]. Copenhagen: Danske Studerendes Fællesråd.

Thomas, C. 1999. *Female Forms: Experiencing and Understanding Disability.* Buckingham: Open University Press.

Vedeler, J.S. 2009. When Benefits Become Barriers: The Significance of Welfare Services on Transition into Employment in Norway. *ALTER European Journal of Disability Research*, 3: 63–81.

White Paper 40. 2002–2003. Dismantling of Disability Barriers: Strategies, Objectives and Political Measures Aimed at Persons with Reduced Functional Ability. Oslo: Ministry of Health.

8 What about landscape in time-geography?

The role of the landscape concept in Torsten Hägerstrand's thinking

Tomas Germundsson and Carl-Johan Sanglert

Introduction

The importance of the landscape concept in the thinking of Torsten Hägerstrand and in his development of time-geography is already well known. On numerous occasions, Hägerstrand makes reference to the experience of being in a landscape and to the landscape as an analytical concept (see Lenntorp 2011; Ellegård and Svedin 2012; Olwig 2017). Nevertheless, the landscape concept has received relatively little attention in the overall development of time-geography and in adaptation of Hägerstrand's thinking in general. This could perhaps be because the landscape is often taken for granted in time-geography representations and is therefore reduced to a mere facet, or static representation, serving as a backdrop to the time-geography notation. Despite this, Hägerstrand himself kept returning to the landscape concept throughout his career, exploring its potential as a basic geographical notion. This relationship with the landscape was probably also quite productive for his geographical thinking, yet it seems as though he never really came to terms with it. He once called landscape the "problem child" of geography (Hägerstrand 1984/1991: 52) and constantly tried to find alternative or complementary concepts to express his ideas about the wholeness he found in a landscape. It can appear paradoxical that Hägerstrand discussed landscape as the study of a contextual totality and simultaneously developed the graphic notation apparatus of time-geography, which often represents individuals as trajectories moving in a seemingly abstract time-space. However, given Hägerstrand's naturalist way of thinking about the landscape as a whole, with all its substances, things, bodies and other matters moving through time, it is perhaps not that surprising that he thought about tools for analyzing the conditions in which this wholeness is constituted and why it changes. Time-geography is a method and a conceptualization for doing just that. Its holistic perspective includes not only tangible "stuff", but also minds, social relations, political influences, administrative rules and other societal matters (see Hägerstrand 1991b).

Based on these arguments, the landscape concept could be put to better use in time-geography, and time-geography could play an important role in landscape studies. Although Hägerstrand rarely engaged in the intra-disciplinary debate on landscape as it developed after the 1970s, with its keen focus on the symbolic, metaphorical, cultural, political and experiential dimensions of landscape (Wylie 2007), many of the issues he raised are relevant to several themes within post-1970s landscape studies. It is of course impossible to cover exhaustively the role of landscape in Hägerstrand's thinking and its connection to broader landscape studies. Our aim in this chapter is rather to assess how Hägerstrand's ideas on landscape could be beneficial in current landscape studies. We do this by reviewing what we believe are some important phases in integration of landscape and time-geography in Hägerstrand's writings. We mainly follow chronological order, i.e., we follow the trajectory of the landscape concept in Hägerstrand's writings using a selection of his work (particularly Hägerstrand 1983, 1984/1991, 1991a, 1993, 2009). Some of these critical texts are in Swedish, so in this chapter we reveal aspects of Hägerstrand's writings on landscape that have not been accessible previously to an international audience. We also present a summary of the "landscape of Hägerstrand", as we understand it. Finally, we present some thoughts on how time-geography and landscape studies can benefit from each other in the future. First, however, we provide a brief background linking time-geography to the landscape concept in general.

Why landscape in time-geography?

The concept of landscape has been central in geography since its creation as an academic discipline, intertwined mainly with regional geography, human-environment issues, and historical studies of change. However, the view on landscape as a field of study has varied over time, due to shifting theoretical and methodological trends within the subject area. The 1980s saw the development of critically oriented landscape geography, which was influenced by contemporary theoretical developments in the social sciences and humanities. In the late 1980s and early 1990s, the discussion came to deal with the landscape as an arena for social power, expressed e.g., in the landscape's symbolic and aesthetic dimensions (see Cosgrove 1984/1998; Cosgrove and Daniels 1988; Duncan and Duncan 2001; Benediktsson 2007). Many studies also portray the landscape as a place for cultural, social and political action (Olwig 1996, 2002a; Mitchell 1996, 2008; Mels 1999; Widgren 2004). The perspective thus moved from a perception of a landscape or region as an interplay between man and nature (Sauer 1925; see Wylie 2007: 17–54) to increasingly addressing how the landscape is shaped and controlled by social processes.

Today, landscape geography encompasses several different and sometimes conflicting perspectives. These range from exploring the landscape as physical space and the ways in which we as humans relate to our environment, to investigations of the visual meaning of landscape, its inherent discursive and ideological tensions, and its role in a place-oriented understanding of community and daily life, including issues of power, inclusion, exclusion and justice (Setten 2002; Henderson 2003; Olwig 2005; Cosgrove 2006; Olwig and Mitchell 2007; Mitchell 2008; see Wylie 2007: 96–138).

Several aspects of Hägerstrand's thinking on time-geography relate to the themes within current landscape geography listed above, although this is generally not explicitly expressed in his writings. Fruitful dialogue between landscape and time-geography is sometimes impeded by the criticism of time-geography expressed by different theoretical and philosophical voices. We briefly describe some of that criticism before presenting the connection between landscape and time-geography in Hägerstrand's own writings.

Time-geography has been described as non-theoretical (Åquist 1991: 71–72; Lenntorp 1999: 155–156; 2004: 224; Wihlborg 2005, 2011) and as a reductionist way of representing the environment and the objects and human beings in it (see Rose 1993: 25ff; Sui 2012: 6; Olwig 2017). Swedish geographer Gunnar Olsson argues, e.g., that its trajectories and prisms are unable to accommodate the hopes and feelings of human beings, by focusing on things taken for granted (Olsson 1998: 128). The humanistic geographer Anne Buttimer also identified these problems in time-geography and communicated them to Torsten Hägerstrand (see Hägerstrand 1983). Although this critique is sound, it tends to emphasize the methodological and representational aspects of time-geography. According to Allan Pred, this view has at times been fostered even by time-geographers:

> It is therefore regrettable that all too many time-geography scholars have been content to ignore the underlying dialectics of their framework, to implicitly or explicitly treat coupling and authority constraints as if they were self-materializing givens, and thereby to mislead or dissatisfy many of their readers.
>
> (Pred 1981: 18)

Sui later noted:

> Although he [Hägerstrand] presented time-geography as a socio-environmental web model, his emphasis on corporeality, contextuality, and collateral processes has often been interpreted in a much narrower sense.
>
> (Sui 2012: 7)

Commenting on Hägerstrand's own development of time-geography, Sui also states that the "extended version of time-geography is more matter-realistic than people-based" (Sui 2012: 7) and argues that Hägerstrand seemed to be mainly interested in the corporeal entities and materiality found at the meso-level of the everyday (Sui 2012: 7). This interpretation, which reflects an overall theoretical movement in the social sciences towards focusing on the subject and discursive dimensions, misses the fact that Hägerstrand's matter-realistic aspirations in his later works explicitly focused on the human condition of always existing in a constantly moving "now" (see Hägerstrand 2009: 225ff). It is true that Hägerstrand displays a materialistic worldview and many of his texts deal with the corporeal and the material conditions of being in a landscape, he nevertheless clearly saw the need to incorporate not only the material body, but also the inner world of the subject (Hägerstrand 2000, 2009).

Furthermore, while Hägerstrand in many ways adopted a modern "scientific" perception of the landscape, as reflected in how he dealt with it in his time-geography analyses (Olwig 2017), one of his main ambitions with time-geography was to challenge modern spatial thinking. He was critical of the increasing specialization in Swedish politics and administration that he perceived in the late 20th century (Ellegård and Svedin 2012). He was also critical of much of modern Swedish planning because of its lack of understanding of a historical wholeness in a landscape and because he foresaw that this kind of sectoral, specialist-led planning with its tunnel vision would have unforeseen (and generally unpleasant) consequences for the future (Hägerstrand 1984/1991; see Hägerstrand 1993). Time-geography emerged as a formalized, condensed and abstract way of presenting why this was the case, by redirecting the analysis to a situated context.

More generally, it could be said that the intention behind time-geography was triggered by the need to analyze historically occurring situations, i.e., landscapes (Hägerstrand 1988). It could therefore be seen as a method based on a comprehensive theoretical framework, although this framework is not always easy to distinguish. Throughout his career, Hägerstrand gave hints and glimpses of such a framework, but his constant involvement in official investigations, planning issues and academic tours led to him postpone elaboration of a more coherent theoretical perspective on the problems time-geography sought to solve until the end of his career and life (Hägerstrand 2009).

In the next section, we give a condensed historiography of Hägerstrand's ideas on landscape in time-geography as revealed throughout his writings.

Into geography

The importance of the landscape perspective in the development of time-geography and in Hägerstrand's overall intellectual development, is reflected in his recollections of his early years at university and his following career as a researcher. When studying geography in the 1940s, Hägerstrand greatly enjoyed geomorphology, climatology and field mapping. Classification was part of geography, but he had one fervent question concerning the discipline:

> It was mostly great fun to be in charge of the weather station. . . . I think I can still classify clouds, if need be. . . . But how was everything held together intellectually? The answer is simple . . . it was not.
>
> (Hägerstrand 1983: 243)

Academic geography obviously appeared to Hägerstrand not as a realm of ideas or a perspective, but as an endless array of encyclopedic data where, for instance, "lectures in regional geography were abominably boring" (Hägerstrand 1983: 244). A fundamental problem for Hägerstrand was the lack of a processual perspective in the geography he encountered. This prompted him to abandon the task of writing a descriptive regional geography of a part of southern Sweden that his supervisor had allocated and instead to focus on migration within, to and from that area in a 100-year period, 1840–1940. This did not involve analyzing anonymous statistics, but rather excavating (from church registers) the individual biographies of around 10,000 people, unraveling their movements, differing livelihood positions and migration patterns (Hägerstrand 1983). In hindsight, it is not difficult to see traces of Hägerstrand's migration studies in the 1940s in his creation of time-geography in the early 1970s. This also shows the importance of context: in his 1940s work he saw that "life-line by life-line is interlaced over the area like the voices of an infinitely complex fugue, repeatedly the same but never quite the same" (Hägerstrand 1983: 245). His archive studies were complemented by field studies, and when viewed in tandem, the combination of data and actual places made a profound impression on Hägerstrand:

> we[1] visited every glade where somebody had tried to make a living. . . . This unusual form of exploration in time and space produced a general world-picture in my mind which today perhaps one would call "holographic". It is not a special way of formulating problems. It is a special way of forming an image before any questions at all can be asked or answers sought. The arduous acquisition of this image is my single most essential experience as an adult geographer. Almost everything I have done since is somehow extrapolated from it.
>
> (Hägerstrand 1983: 245).

In the early 1970s, after defending his thesis on innovation studies in 1953 and dealing with a number of regional policy issues in Sweden during the 1960s, the time had come for Hägerstrand to start formulating what would become known as time-geography. At the time, he was involved in research on regional development and planning, urbanization, and future studies. Hägerstrand claimed that time-geography was not so much the result of his research at the time, but an effort to express a worldview he had been thinking about for a long time (Sollbe 1991: 213). In a collection of essays during the 1970s, he presented parts of this worldview in what eventually grew into the more coherent conceptualization "time-geography" (Hägerstrand 1970a, 1970b, 1975b, 1982, 1985). Not surprisingly, the broad academic and general acknowledgement of Hägerstrand's work focused on the groundbreaking and visually striking time-space diagrams and the trajectories therein (Brauer and Dymitrow 2017). The academic geographical community rapidly became acquainted with, and fascinated by, the graphic representation of individual and bundled trajectories in time-space, of different kinds of time-space restrictions, and of possibilities of movement.

The content and concept of time-geography was in many ways refined during the 1970s and 1980s, for instance through a more consistent worldview, but also as a more or less formalized research program (see Hägerstrand 1991a: 135). In this process, time-geography often appeared to be a methodological program rather than a theoretical approach, or a method for geographical observation rather than explanation (Hägerstrand 2009: 108; see Lenntorp 1999). For Hägerstrand himself, however, there always seemed to be a close relationship between the models and the understanding of a concrete empirical situation. Time-geography in his own writings was almost always developed in relation to actual circumstances, and many of Hägerstrand's works on time-geography concepts were written in response to, or as a critique of, how different fields of research dealt with contemporary problems, such as regional science (Hägerstrand 1970b), future studies (Hägerstrand 1972, 1975a), planning (Hägerstrand and Lenntorp 1974), demography (Guteland et al. 1974), transport research (Ellegård et al. 1977), and later the integration of nature and society (Hägerstrand 1976, 1984/1991, 1993). For instance, in Hägerstrand's planning research, one of the main areas in which time-geography was developed, there was a very conscious effort to counter specific sectoral tendencies in practical planning and in academic research (Hägerstrand 1972, 1975b; see Hägerstrand 2009: 62, 134). The time-geography notation apparatus as introduced by Hägerstrand did not have the prime aim of constructing a four-dimensional notation system *per se*, we would argue, but rather of forming a conceptual model as part of a much broader and general perspective:

> The central theme in my efforts in planning has been a constant fight against the negative influences on life and landscape of the functionally sectorized "system society".
>
> (Hägerstrand 1983: 253)

On time-geography at large, Hägerstrand wrote:

> For myself at least I have managed to unite within one frame of thought all the various matters I have dealt with earlier: settlement, migration, social communication, diffusion, domain structure, and impact of technology. There is also the beginning of an answer to my old quest for the purpose and form of the "regional" approach to geography and the placing of man in nature.
>
> (Hägerstrand 1983: 254)

For Hägerstrand, the landscape thus functioned as a unifying framework in which almost any kind of question or theme could be studied, transcending the established boundaries of academic disciplines and branches of administration. This line of thought can be traced throughout his writings but took different forms over the years, partly due to the specific questions he was working on at the time, but also perhaps as a result of gradual refinement of concepts and theoretical considerations.

Hägerstrand and the landscape

Time-geography in the landscape

In Hägerstrand's efforts to formulate a broader and more comprehensive framework for the time-geography approach, the landscape served as an important point of reference and from time to time he critically confronted the landscape concept. One such occasion is the memorial lecture to Carl Sauer in 1984, in which Hägerstrand discussed "The landscape as overlapping neighbourhoods" (Hägerstrand 1984/1991). His elaboration of the landscape concept in this lecture included comments on three prominent scholars: Carl Sauer himself, the Finnish geographer Gabriel Granö and the American geographer Richard Hartshorne.

Granö is acknowledged for including all the human's senses when trying to find a scientific method for objectively measuring the total character of the environment (see Granö 1929). However, while Granö incorporates human beings in the landscape, according to Hägerstrand he "draws a sharp line between body and mind" (Hägerstrand 1984/1991: 53), which Hägerstrand considered quite unsatisfactory in a landscape context.

Hägerstrand also sympathetically acknowledged Carl Sauer's endeavor to understand the ways in which the "technological interventions in the physical world and its life has become the crisis of his [man's; referring to humanity's] survival and that of its co-inhabitants" (Hägerstrand 1984/1991: 52). Yet Hägerstrand noted that Sauer in his work "saw the human mind represented – by the broad concept of culture – as a force apart from the landscape itself" (1984/1991: 53). Hägerstrand disapproved of this conceptualization.

The discussion on Granö and Sauer clearly shows that Hägerstrand's landscape conceptualization included the human mind, inseparable from the body, an issue he would return to in several later writings (see Hägerstrand 2004, 2009).

In his memorial lecture Hägerstrand also mentioned Richard Hartshorne's famous dismissal of the landscape concept. Hartshorne believed it had nothing explanatory to say about the togetherness or the wholeness of an area or a region; it just showed a random juxtaposition:

> the fact that the human mind has a unit impression of a collection of things does not prove for a moment that they have in themselves any relations to each other, other than juxtaposition.
>
> (Hartshorne in Hägerstrand 1984/1991: 51)

Hägerstrand strongly opposed this view of juxtaposition as an insignificant relation, arguing that it is exactly this casual "juxtaposition of a mixed assortment of entities . . . which has called forth a need for a word that communicates these characteristics" (Hägerstrand 1984/1991: 51). For Hägerstrand, "landscape" was a key concept and, in spite of the difficulties, of "fundamental scientific importance" in geography (1984/1991: 51) because of its capacity to bring space and time together in a progressive way. It is interesting to note that, when trying to find a common denominator for the fundamental realms under which humans act in the world, Hägerstrand described these as "our personal everyday activities, the real events which we tell stories about, and the various technical applications of scientific knowledge" (1984/1991: 52). He did not use the concepts of time and space, but instead place, room and duration:

> They [the fundamental realms above] all need a place or places to be, they need *room* at their disposal over a sufficiently long *duration*.
>
> (Hägerstrand 1984/1991: 52)

The use of these relational concepts suggested a connection to traditional regional geography and also called for a more contextual approach both to time and space:

> we ought to understand what the unbroken persistence of beings from birth to death means to their mutual relations.
>
> (Hägerstrand 1984/1991: 54)

In our view, this is very much a call for a time-geography approach to landscape studies, although Hägerstrand did not mention time-geography in his lecture. However, in the lecture he acknowledged that the landscape concept is essential in geography, but also problematic. A full empirical description is beyond reach, and therefore Hägerstrand suggested a method that calls for quite a formal approach (for which Sauer is posthumously excused in the memorial lecture). Hägerstrand maintained that this formal approach is necessary to come to grips with the environmental problems that arise from modern society and that are generally not recognized until it is too late.

Man, nature and society

By the mid-1980s, Hägerstrand's development of a contextualized time-geography seemed to move into a phase where its potential to scrutinize scientific problems common to the natural and social sciences became explicit, in particular environmental problems and problems within physical planning. For Hägerstrand, these were largely problems deriving from the disciplinary divisions and specializations within academia, but also within government administrations, rendering these institutions unable to cope with the complexity of everyday life. While Hägerstrand most certainly saw the potential of the natural sciences in research on environmental issues, he maintained that a better ability to work with a contextual wholeness was crucial to solving the looming problems stemming from man's current relations to nature (see Carlestam 1991: 14ff). Hägerstrand's own approach was mainly materialistic and naturalistic, focusing on the bodily existence and interaction between different entities in a given environment. This continually led him to the problematic but unescapable landscape concept:

> The best approximation we have in geography to a concept capable of grasping the momentary thereness and relative location of all continuants is *landscape* . . . rules and regulations receive their reality by being understood and respected as real by humans, and they are therefore present in the landscape as much as are the things we can see and touch.
>
> (Hägerstrand 1982: 325–326)

This indicates that landscape in Hägerstrand's thoughts was not meant to be understood solely as a physical concept, but rather as the meeting point between the social and the physical. Yet Hägerstrand's focus on the material aspects of the nature–society interaction, as represented in his time-geography diagrams, appeared to overlook giving a voice to the living human beings therein. When challenged by Anne Buttimer, an astonished Hägerstrand had to admit that his time-geography apparatus could be seen as depicting a bloodless, mechanical "*danse macabre*",[2] which he claims struck him like a bolt of lightning (Hägerstrand 1983). The period of meeting and working with Anne Buttimer was undoubtedly an important turning point for Hägerstrand and his geographical thinking, not least, we would argue, for bringing in personal and subjective experiences of places and landscapes in his scientific endeavors, but also for conveying a humanistic legacy within geography, philosophy and literature, which was to be reflected in his later writings (see Hägerstrand 2009).

Such a turn can be noted in the article "Time-geography" (Hägerstrand 1985), in which Hägerstrand focused on what he labeled the "society-nature-technology trialectics" as a theoretical basis. Hägerstrand used the "three worlds" of Karl Popper and John Eccles – consisting of the physical objects, mental states (mind), and the cultural products created by the human mind (immaterial and material) – as an inspiration. These three worlds are so connected that they mutually condition events and relations within each other (Hägerstrand 1985: 194). To proceed, Hägerstrand "as a geographer, accustomed to viewing the world in terms of landscapes" introduced a first step to analyze the problem of society–nature–technology trialectics in a contextual manner, namely to "define a bounded region and accept what is found in it . . . as the given universe" (Hägerstrand 1985: 195; see Hägerstrand 1993). This development of the landscape concept relates to the emergence of what Hägerstrand later termed a socio-ecological and topo-ecological view on landscapes and spatial relations (Hägerstrand 2004: 323). This could largely be attributed to his engagement in a contemporary environmental discourse, while inclusion of the mind and the human subject might be seen as a response to the critique of time-geography advanced by other geographers with whom he had dialogue (e.g. Derek Gregory, Christian van Paasen and Anne Buttimer).

Landscape, ecology and process

In the early 1990s, an ambition to dissolve the superficial border between nature and society led Hägerstrand to think more explicitly about scientific ways to approach a concrete all-inclusive landscape. Although critical of much of the geographical tradition, Hägerstrand often celebrated the landscape concept of the "good" synthesizing regional geography, as a contrast to a contemporary sectoral discussion on, e.g., spatial planning, nature conservation and environmental issues. In several writings, he emphasized the central function of the landscape as an empirical starting point, in one case described as "a history book . . . a source of knowledge of both nature's and culture's development place by place" (Hägerstrand 1988, translated from reprint in Carlestam and Sollbe 1991: 39). At the same time, Hägerstrand was critical of the way in which historians, but also geographers, described spatial cross-sections of the earth's surface (Hägerstrand 1984/1991: 54) and tended to deal with the landscape and spatial relations in a surface-like manner. He also criticized the tendency to study landscapes through what he called "discrete leaps" between different periods or stages, rather than studying continuous flows (Hägerstrand 1993: 27; see Carlstein 1982: 23). Moreover, he was critical of much of the administration and practical management of the landscape, and on several occasions advocated an integrated approach, e.g., in nature conservation and cultural landscape heritage. In a less formal paper, Hägerstrand wrote that the administration of nature and culture preservation would work much better if the landscape was looked upon and treated as a garden – a wholeness woven together by activities and a sense of care – and not as administrative domains of nature reserves and cultural heritage, treated from scientifically divided perspectives (Hägerstrand 1988).

Hägerstrand argued that, not least from an ecological and environmental perspective, the landscape should be studied as a spatial wholeness, as a total room. He wrote that he wanted to utilize the landscape concept to develop a *hopfogningslära* (science of amalgamation) (Hägerstrand 1986), i.e., an understanding of the conditions under which our reality takes shape and how we can comprehend the world around us as a relational set-up of "things beyond things" (Hägerstrand 1991a; see Williams 1953: 457ff).

From 1990 onwards, Hägerstrand worked on generating such a path forward in an ecological context, as he was worried about the consequences of modern society's unsustainable way of handling resources and the environment. One of the conceptual strains, long present in modern thinking, that he tried to resolve in his work was the division between nature and society.

In a longer article in 1993, following up on these issues and on the critique of the current spatial planning and resource management in times of structural change within society, business and government planning, Hägerstrand used the landscape perspective to outline the kind of knowledge needed for fostering more sustainable development (Hägerstrand 1993). That article, entitled "Society and nature" (Swedish, "Samhälle och natur"), took its point of departure in the idea that the long-held social contract as a base for a democratic development has to be supplemented with a "nature contract", an idea launched by the Swedish historian of ideas Sverker Sörlin (Sörlin 1991). To Hägerstrand the geographer, such an ambition had to be investigated and developed by studying concrete real-world settings where natural and social processes unfold simultaneously and are intertwined, and where the adverse effects of ignoring nature's limited capacity to cope with man's exploitation of resources and stress on ecological systems actually play out. For Hägerstrand, the only setting that could be seen as fully representative of all present entities and processes was a piece of populated landscape. In his 1993 article, Hägerstrand thus introduced a methodological principle of studying the landscape as a common space that is necessarily shared by all entities present in that landscape during a certain period. The purpose was not to perform any local or regional landscape studies *per se*, but to establish an approach that could connect processes on different scales in a concrete landscape evolving over time. In Hägerstrand's vision, the landscape included not only that which is visible, but also all other entities within the defined borders of the study area, including everything moving in and out of the area during the study period (Hägerstrand 1993: 26). Such a perspective certainly had its methodological and epistemological implications. One of Hägerstrand's means to tackle that challenge was to introduce a dual landscape concept, partly in analogy with the earlier distinction between the inner and outer world (Hägerstrand 1990; see Lenntorp 1976: 14; Mårtensson 1979). Hägerstrand (1988) had hinted at this, but now developed it further. The concepts were labeled "the view landscape" and "the passage-of-event landscape" (in Swedish *utsiktslandskap* and *förloppslandskap*), in the following called the "viewscape" and the "processual landscape".[3]

The viewscape is the landscape we can observe, the sensibly conceived landscape; it is a concrete reality but, due to our human condition, we can only perceive and communicate about it from an egocentric perspective (Hägerstrand 1993: 26; see Hägerstrand 2009: 264). It is the landscape we talk about in terms of administration, management, conservation, etc. The perceived viewscape, however, has its concealed conditions, processes, flows and constraints, and therefore a complementary approach and representation to capture its dynamics is needed. In addition to the viewscape, Hägerstrand therefore introduced the concept of the processual landscape. Like the orthogonal map, it is an "all-seeing" landscape conception, but unlike the map it does not emphasize the visual or the static. Drawing on Hägerstrand's imagining of the totality in any landscape, the processual landscape consists of everything that is present: remains of all past activities and all things on the way somewhere in space and time under certain conditions (Hägerstrand 1993: 26). In elaborating on the concept and approach of the processual landscape, Hägerstrand explicitly engaged time-geography and illustrated how it could work as a frame of thought for how the physical and material landscape is integrated with social institutions and transactions, e.g., domain structure and aspects earlier discussed by Hägerstrand as "pockets of local order" or time-space domains (Hägerstrand 1970b: 18; 1985: 207ff).

In the processual landscape everything is included, constantly occupies space, and passes through time. Transformation is continuously ongoing through permutation and succession, which encompasses both nature's and society's competition for space. Studies of the processual landscape therefore become a matter of observing and interpreting the physical "power exercises" of different phenomena in relation to each other (Hägerstrand 1993: 27). While the viewscape is the perceived concrete reality around us, and therefore more in line with a conventional meaning of landscape, due to its wholeness the processual landscape should preferably be the object for management (Hägerstrand 1993: 40ff). In its capacity of representing the all-seeing eye conceptually it must, however, also remain more of an imagined phenomenon, because of its complete totality and full extent in time and space (Hägerstrand 2009: 268). Needless to say, studies of the processual landscape must therefore always be practically based on selections and prioritizations. However, these methodological processes are then grounded in efforts to go beyond the conventional political, administrative and scientific sectors and focus on problems occurring in a concretely situated environment – a landscape – and what could be done to solve them.

As demonstrated, Hägerstrand sought a formalistic approach that could explain the conditions under which the complexity and relational togetherness of the landscape worked as constant ongoing processes. Time-geography, working as a "situational science" (see Asplund 1983: 189), became a tool for Hägerstrand in this pursuit, and acted as a kind of skeleton for his use of the landscape concept. To put some flesh on the bones, Hägerstrand "tested" the landscape concept in different relations to time-geography principles, as reflected in his creation of new concepts like the "viewscape" and "processual landscape". Another example is the concept of *andskap* – a word play on the Swedish *landskap* (landscape) and *ande* (spirit) perhaps best translated as "mindscape", but possibly also "spiritscape" (Hägerstrand 1991b). This indicates the insoluble consolidation of physical matter, the body, and the mind in Hägerstrand's landscape conception. It should also be noted that it was not only spatial or place-bound phenomena which were comprehended as having relational qualities, but also time. Depending on the context, Hägerstrand viewed time both as a constantly ticking clock and in a more relational understanding as episodes, rhythms and temporalities (Hägerstrand 2000: 122; see Sui 2012).

Landscape as the totalizing concept of geography

As demonstrated above, Hägerstrand had an ambivalent attitude to the landscape concept throughout his career and even in his last work, the posthumously published book *Tillvarovären* (tentatively *The Fabric of Existence*) (Hägerstrand 2009). In this work, Hägerstrand tried to summarize his worldview and also – in a notably longer format than in his earlier writings – to give a more comprehensive picture of the relationship between society and nature.

In the book, Hägerstrand addressed a number of ontological and philosophical questions regarding human relations to time and space. Although initially actually dismissing the need for a fixed concept of landscape (Hägerstrand 2009: 59–60), later in the book he eventually returned to the concepts of the viewscape and the processual landscape, in order to describe the relationship between the perceived landscape as a socially conditioned idea and the external landscape as a total, but ungraspable, material reality (Hägerstrand 2009: 264, 267).

The book reflects the fundamental and enduring role of time-geography in Hägerstrand's own geographical imagination. Despite the criticism directed at time-geography over the years, partly by Hägerstrand himself, he never abandoned the concept. In fact, in *The Fabric of Existence* he indicated that the premises of time-geography were critical for his worldview. Consequently, and in line with much of his earlier work, time-geography emerged as more or less fundamental to Hägerstrand's conceptualization and utilization of landscape as a scientific concept.

Some lines of reasoning in *The Fabric of Existence* demonstrate how, in our view, Hägerstrand's development of a comprehensive worldview could be conceived as an argument for exploring time-geography's potential in land-scape studies. Elaborating on the thoughts and the reasoning underpinning his worldview, Hägerstrand noted that the contact, or rather encounter, in time-space between any two entities is the prime elementary event in the world, in the fabric of existence, irrespective of context and scale (Hägerstrand 2009: 69–70). This could be anything from the composition of molecules to the ferti-lization of the egg, or to planetary and celestial movements. The fundamental importance of encountering implies other aspects of existence such as detach-ing, but also *shortage* of encounters (Hägerstrand 2009: 40, 71, 136), as well as qualitative differences in terms of distance between entities, both physically in material space and negotiated through social structure (Hägerstrand 2009: 49). Hägerstrand typically invented new words when conventional words would not suffice. For example, he used "side-by-side-ness" (Swedish, *bredvidvartan-nathet*) (Hägerstrand 2009: 158) to characterize the furnishing of the mundane world and also in order to describe the relational conditions in the fabric of existence (see Hägerstrand 2009: 24, 29).

Hägerstrand explicitly stated that his goal with these ontological and theoreti-cal elaborations was to form a human ecology. It is clear that he wanted to give equal weight to both words in the concept and that he thereby sought to actively transgress the borders between the social and the material worlds (Hägerstrand 2009: 62). For Hägerstrand, the corporeal existence in time-space was the obvi-ous medium, or interface, between these two worlds that he called the inner and outer world (Hägerstrand 2009: 228ff). Picking up on the concepts of "path" and "continuant" from the broader framework of time-geography, the concept of the body as such an interface was graphically expressed in the four-field model of the "now-point" (Hägerstrand 2009: 229) (Figure 8.1).

The illustration in Figure 8.1 demonstrates how Hägerstrand wanted to include all things – including the sentient human being – in a spatial wholeness, a total room (*allrum* in Swedish), which was a term and an idea he borrowed from the phe-nomenologist Husserl (*Gesamtraum*) (Hägerstrand 2009: 84). Husserl's thoughts helped Hägerstrand elaborate on the existential issue of human beings' own "thingness", their "I-bodies" (*Ichleib* from Husserl) in a total space (Hägerstrand 2009: 82). This phenomenological perspective and the notion of the now-point also deepened the landscape perspective, tying the viewscape and the processual landscape together, and could be seen as a continuation of the humanistic perspec-tives raised in earlier texts.

Hägerstrand suggested another concept closely related to the viewscape and the processual landscape, namely, the "resulting space" (Swedish, *utfallsrum*), the temporary product of the processual landscape and the interaction of differ-ent bodies therein (Hägerstrand 2009: 229). This resulting temporal landscape may be perceived as a frontline (see Hägerstrand 2009: 59–60, 269, 271; see also Lenntorp 2011), a conglomerate of individual now-points, in a continuous move forward. This temporal landscape is both the result and the prerequisite of the actions making up the very fabric of the processual landscape (Hägerstrand 2009: 65, 78).

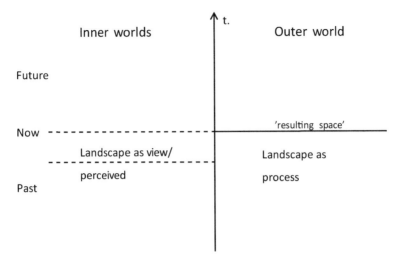

Figure 8.1 Relationship between the inner and outer world and between past and present,
all coming together in the "now-point" (Swedish, *nu-punkten*) (based on
figure 11 in Hägerstrand 2009). The "now-line" in the outer world is sharp
and moving, whereas the distinction between past and future in the inner
world is more diffuse, hence the two dotted lines framing the inner "now"
(see Hägerstrand 2009: 228–229). In our interpretation, the outer processual
landscape (lower right) is in a constant process of generating the resulting
space (Swedish, *utfallsrum*) along the now-line, while moving further into
future possibilities (upper right). The inner experienced landscapes (lower
left) have a more diffuse now-line ("now-zone") that is permeable in relation
to the imagined landscapes (upper left).

Through these references to *The Fabric of Existence*, we show that Torsten
Hägerstrand, in his efforts to express an elaborated and coherent conceptualization of the material conditions in which nature and society change and develop,
focused on elementary events and encounters in this topo-ecological framework
and constantly discussed how we – as human beings – can make meaning and
act in such a topo-ecology. Although not explicitly claiming that landscape had
a prior status in this framework, we believe that Hägerstrand's firm focus on the
"side-by-side-ness" of things in a concrete setting, coupled with his constant
eagerness to discuss this from the perspective of a human being, make his ideas
appropriate for current landscape geography. In the next section, we summarize
some aspects underpinning this.

The landscape of Hägerstrand

Landscape undoubtedly played a vital part in Hägerstrand's thinking and, we
would argue, landscape and time-geography had a dialectical relationship in

his work. While experiences and problems from real-world landscapes seem to have been important sources of inspiration in the development process of time-geography – providing an analytical framework to study these problems – Hägerstrand appears to have been unable to relinquish the messy landscape concept, constantly returning to it with new questions, not least informed by his development of time-geography.

Is it then possible to distinguish a "landscape of Hägerstrand"? Although we do not consider it meaningful to try to formulate any final or essential comprehension of Hägerstrand's landscape thinking, there are a number of fundamental and recurring themes in his writings that could be fruitfully pieced together for future discussion. To demonstrate this, we argue that Hägerstrand's theoretical framework for understanding the landscape consisted of three different ontological levels or spheres.

The basic sphere is the *material circumstances of existence* in the world. This theme was not common in his earlier texts, but was elaborated on at the beginning of his posthumous book *The Fabric of Existence*, where Hägerstrand first characterized the material totality's "thingness", the tendency of material matters to form a grain structure (Swedish, *kornighet*) or a thing structure (Hägerstrand 2009: 77). In this, the composition of a situation – a landscape – is constituted by the object's "side-by-side-ness", relating to the earlier description of the material world as relational build-up of "things beyond things" (Hägerstrand 1991a) existing in a "complementary space" that is not void but constituted by its relation to the things in it (Hägerstrand 2009: 87). This materiality sets the preconditions for the fundamental elementary event of "encounter" (things bumping into each other), without which there would be no world.

This leads to the second sphere, namely, *the landscape process*, the time-space volume (Hägerstrand 1993: 36ff), in which processes, actions and events gradually unfold and take place in the fabric of existence, like patterns of individual pathways formed by the local structure of domains (Hägerstrand 1970b; see Hägerstrand 1985: 207ff). There, the concepts of time and space are not to be understood as derived from an abstract coordinate system, but from material encounter configurations (Hägerstrand 2009: 80). The processual landscape is in constant flux, but also holds a certain kind of inertia – a resistance to change – by its historically built-up structure. The resulting landscape could at any moment in time be perceived as the temporal imprint of the paths forming the treads in the landscape weave (Hägerstrand 1974: 90). However, while landscape can definitely say something about common contemporary circumstances, things present in a particular landscape tend to be in different phases of birth, life and death and move simultaneously, but in different phases (Hägerstrand 1988). Landscapes should therefore be perceived as relational heterogeneous contexts rather than as homogeneous representations of a certain phase. In Hägerstrand's complex of landscape concepts, this temporal result serves as an interface between the processual landscape and the viewscape.

The viewscape is the third sphere, constituting the total physical landscape we can observe and communicate about. It is thus inescapably *simultaneously a material and a mentally, culturally and socially constructed landscape*, whether it is made conceivable through art, academia, administration or simply aesthetic experience (Hägerstrand 1988; Hägerstrand 2009: 90).

This link between the physical and the mental, or cognitive, dimension of reality, or as Hägerstrand expressed it, between the outer and inner world (Hägerstrand 2009: 228ff), is in fact a central point not only in what we might call his landscape view, but also in the time-geography perspective. The ambition to link the physical and social is clearly reflected in the wording of the main restrictions of "authority", "coupling", and "capacity" that Hägerstrand believed were largely the reason for the emergence of different domains (Hägerstrand 1970b) or pockets of local order (Hägerstrand 1985). As shown above, the link between the social and physical dimensions of existence is also found in the model of the so-called now-point, the point at which both the past and future and the inner and outer world constantly converge.

In this brief summary, we have shown how Hägerstrand's materialistic, and sometimes almost mechanistic, worldview became a tool for him when scrutinizing humans' living conditions and possibilities of action and change in their landscape. A constant challenge to him in this pursuit was how to integrate conscious acting individuals in this landscape without representing them in the conventional way in a two- or even three-dimensional space. Rather, he presented them as bound by their own mental, social and bodily restrictions and capacities when acting to find room for existence in a constant ongoing relation to the surrounding landscape of humans, things, and socially given structure.

Moving forward in the landscape of time-geography?

In this analysis, we investigated how the landscape and the landscape concept are integrated in Hägerstrand's work. The aim was to identify how the landscape as a basic geographical concept relates to time-geography as developed by Hägerstrand. While time-geography can obviously and correctly be seen as Hägerstrand's main contribution to geography, it can also be fruitful to discuss it in relation to other geographical concepts and ideas. We chose landscape as a concept, partly because it is more or less explicitly present in much of Hägerstrand's work, and partly because Hägerstrand's geographical imagination at large, not least in relation to how he developed time-geography, was founded on his scientific curiosity and ambition to understand and explain humans' complex and composite habitat in a concrete spatial and evolving context.

We uncovered a tight dialectic relationship between time-geography and landscape in Hägerstrand's development of his geographical thinking. This is not a novel insight, e.g., critics and advocates of time-geography and previous reviewers have observed this symbiosis (see Sui 2012). Sui sets time-geography at the heart of Hägerstrand's thinking, whereas we chose a different starting point in the landscape. This gives us reason to look somewhat differently on Sui's observation, that over time Hägerstrand "shifted his perspective of reality from an orthogonal to an oblique view, which is more personal, aesthetic, and subjective", and that "understanding Hägerstrand's shifting view of space from one defined by abstract coordinates, to place, as defined by landscapes, is crucial for fully appreciating Hägerstrand's multiple dimensions of time-geography" (Sui 2012: 8). While the landscape view is certainly crucial for understanding Hägerstrand's ambitions with time-geography, and how it evolved over time, we would argue that such an understanding of place and landscape had long been present in his thoughts (see Olwig 2002b). The observation by Sui reflects a general trend in the social sciences to adopt a subject-oriented research perspective. However, we believe that, rather than shifting the conception of space from one view to another, the development of Hägerstrand's geographical thinking was driven by the challenge of keeping both views alive at the same time, i.e., integrating them into each other, as both are crucial for interpreting the world we live in – the inner and outer world.

It is true that time-geography's mode of representation is often an abstract image of space and time, linked both to the all-seeing eye from above, i.e., the map, and to the perspective of representation from a certain point of view. Olwig (2017: 49–50) argues that time-geography thus demonstrates a "spatial reduction . . . that allowed Hägerstrand to transcend the map and turn it into a four-dimensional pictorial space-time model". However, as Olwig also notes (2017: 61), Hägerstrand obviously recognized the tension between his situated human ecological perspective and the graphic time-geography representation. This did not lead him to abandon the latter, but rather to work on filling the gaps. As we see it, Hägerstrand found reason to utilize the modern comprehension of time and space, but also to challenge it. His endeavor to comprehend the human environment as a whole – a totality – led him to look for a form of representation that worked outside the conventional scale and that provided a double perspective, from a certain standpoint (the human perspective) and from above (a modern spatial perspective) (see Ellegård and Svedin 2012). He therefore never abandoned his ambition to represent or think about the landscape in a perspectival and cartographic way, or to understand the world as a reality that can be denominated in terms of things, packages, grain structure, encounters, etc. (Hägerstrand 2009). He considered this way of conceptualizing the landscape necessary to complement the limited and situated perception of the world we have as humans, as a means to make it manageable in a more systematic way, and thus change it in a desirable way. Hägerstrand needed the double concepts of the situated viewscape and the orthogonal processual landscape in order to capture the totality, thereness and togetherness he described in *The Fabric of Existence* and to identify the circumstances in which all organisms can live together on the limited surface of the earth. The task is immense, of course, and, as Hägerstrand himself noted, it is easy to pose the "all-at-once" question, but very unclear how to answer it (Hägerstrand 2009: 241). Investigations must always be piecemeal.

What Hägerstrand had in mind with time-geography was not simply an analytical approach to study human movement in time and space, but also a broadly based conceptual framework for geographical synthesis. From our examination of landscape thinking in Hägerstrand's work, we would add that the synthesis that Hägerstrand sought should not be viewed as achievable simply by advancing and refining time-geography. While Hägerstrand never came to grips with the landscape concept as the consistent scientific concept that he was looking for, his constant return to it reflects some dimension of the kind of synthesis he sought, building on, but also going beyond, time-geography. This opens up fruitful ways forward for future landscape studies, e.g., in furthering the phenomenological path that Hägerstrand hinted at, which today is a lively branch of landscape geography (Tilley 1994; Hannah 2013; Wylie 2013); in including actor-network theory (ANT) and its equating of nature and society in a manner that resembles Hägerstrand's ideas, an approach already used in landscape studies (Qviström 2003, 2015; see Schwanen 2007); or in opening up possible connections to the ontological stratigraphy of critical realism as a basis for landscape studies (Sanglert 2013).

The opportunities are manifold. We hope that this chapter on the relevance of landscape in Hägerstrand's development of time-geography, and also in his geographical imagination, can act as inspiration for further studies.

Notes

1 TH and his by then fiancée and later wife, Britt Lundberg.
2 A medieval artistic genre depicting the inevitable death of all human beings.
3 The Swedish word *utsikt* means view, especially a view from a height, but also more generally: "what a nice view from the window". We chose to use the term "viewscape" because we think that this English construction (which would be very clumsy in Swedish) quite well covers the meaning Hägerstrand gave to the concept.
  Förlopp in Swedish means "passage of events" or maybe "course", but also leans towards "process". *Förloppslandskapet* has previously been translated as "the processual landscape" (Germundsson and Riddersporre 1996), and we believe that the slightly more active dimension of "process" (compared to "passage of events") is more consistent with the meaning Hägerstrand gave to the concept.

References

Åquist, A.-C. 1991. *Tidsgeografi i samspel med samhällsteori*. Lund: Lund University.

Asplund, J. 1983. *Tid, rum, individ och kollektiv*. Stockholm: Liber Förlag.

Benediktsson, K. 2007. "Scenophobia", geography and the aesthetic politics of landscape. *Geografiska Annaler: Series B, Human Geography*, 89: 203–217. DOI: 10.1111/j.1468-0467.2007.00249.x

Brauer, R., and Dymitrow, M. 2017. Human geography and the hinterland: The case of Torsten Hägerstrand's "belated" recognition. *Moravian Geographical Reports*, 25(2): 74–84.

Carlestam, G. 1991. Samtal om verklighetens komplexitet, kunskapens och språkets gränser. In: Carlestam, G. and Sollbe, B. (eds) *Om tidens vidd och tingens ordning: Texter av Torsten Hägerstrand*. Stockholm: Statens råd för byggnadsforskning, pp. 7–18.

Carlstein, T. 1982. *Time Resources, Society, and Ecology*, vol. 1: *Preindustrial Societies*. London: Allen & Unwin.

Cosgrove, D. 1984/1998. *Social Formation and Symbolic Landscape*. Madison: University of Wisconsin Press.

Cosgrove, D. 2006. Modernity, community and the landscape idea. *Journal of Material Culture*, 11: 49–66.

Cosgrove, D. and Daniels, S. 1988. Introduction: Iconography and landscape. In: Cosgrove, D. and Daniels, S. (eds). *The Iconography of Landscape*. Cambridge: Cambridge University Press, pp. 1–10.

Duncan, J., and Duncan, N. 2001. The aestheticization of the politics of landscape preservation. *Annals of the Association of American Geographers*, 91(2): 287–409.

Ellegård, K., and Svedin, U. 2012. Torsten Hägerstrand's time-geography as the cradle of the activity approach in transport geography. *Journal of Transport Geography*, 23: 17–25.

Ellegård, K., Hägerstrand, T. and Lenntorp, B. 1977. Activity organization and the generation of daily travel: Two future alternatives. *Economic Geography*, 53(2): 126–152.

Germundsson, T., and Riddersporre, M. 1994. Förloppslandskap och bevarande. In: *Svensk Geografisk Årsbok 1994*, pp. 144–150.

Granö, J.G. 1929: *Reine Geographie*. Helsingfors: Tilgmann.

Guteland, I., Holmberg, A., Hägerstrand, T., Karlqvist, A., and Rundblad, B. 1974. *The Biography of a People: Past and Future Population Changes in Sweden, Conditions and Consequenses*. Stockholm: Allmänna förlaget.

Hägerstrand, T. 1970a. Tidsanvändning och omgivningsstruktur, *SOU* 1970:14, bilaga [appendix] 4, pp.1–146.

Hägerstrand, T. 1970b. What about people in regional science? *Regional Science Association Papers*, 24: 7–21.

Hägerstrand, T. 1972. Att välja framtid. *SOU 1972:59. Betänkande avgivet av arbetsgruppen för framtidsstudier* (A. Myrdahl, M. Fehrm, M. Frankenhaeuser, T. Hägerstrand, L. Ingelstam, B. Odén, I. Ståhl, A. Engström and Å. Mattsson). (In English: *To Choose a Future*, Stockholm: Svenska institutet, 1974.)

Hägerstrand, T. 1974. Tidsgeografisk beskrivning: Syfte och postulat. *Svensk Geografisk Årsbok 1974*, pp. 86–94.

Hägerstrand, T. 1975a. Kunskap om framtiden ur det förgångna. *Lundaforskare föreläser 7.* Lund: Lund University Press.

Hägerstrand, T. 1975b. Survival and arena: On the life-history of individuals in relation to their geographical environment. *Monadnock*, 49: 9–29.

Hägerstrand, T. 1976. Geography and the study of interaction between nature and society. *Geoforum*, 7(5/6): 329–334.

Hägerstrand, T. 1982. Diorama, path and project. *Tijdschrift voor economische en Sociale geografie*, 73(6): 323–339. (Also published in J. Agnew, D. Livingstone and A. Rogers (eds.), 1996. *Human Geography: An Essential Anthology*. Oxford: Blackwell, pp. 650–674.)

Hägerstrand, T. 1983. In search for the sources of concepts. In: Buttimer, A. (ed.). *The Practice of Geography*. London: Longman, pp. 238–256.

Hägerstrand, T. 1984/1991. The landscape as overlapping neighbourhoods. In: Carlestam, G., and Sollbe, B. (eds). *Om tidens vidd och tingens ordning: Texter av Torsten Hägerstrand*. Stockholm: Statens råd för byggnadsforskning, pp. 47–55. (Earlier unpublished lecture.)

Hägerstrand, T. 1985. Time-geography: Focus on the corporeality of man, society, and environment. In: *The Science and Praxis of Complexity*. Tokyo: United Nations University, pp. 193–216.

Hägerstrand, T. 1986. Den geografiska traditionens kärnområde. *Svensk Geografisk Årsbok 1986*, pp. 38–43.

Hägerstrand, T. 1988. Landet som trädgård. In: Heurling, B. (ed.). *Naturresurser och landskapsomvandling: rapport från ett seminarium om framtiden*. Stockholm: Bostadsdepartementet, pp. 16–55.

Hägerstrand, T. 1990. På väg mot ett integrerat perspektiv. In: Aniansson, B., Kågesson, M., and Svedin, U. (eds). *Individ, samhälle och miljö*. Rapport 90:1. Forskningsrådsnämnden, pp. 79–84.

Hägerstrand, T. 1991a. Tidsgeografi. In: Carlestam, G., and Sollbe, B. (eds). 1991. *Om tidens vidd och tingens ordning: Texter av Torsten Hägerstrand*. Stockholm: Statens råd för byggnadsforskning, pp. 133–142.

Hägerstrand, T. 1991b. Tillkomsten av nationalparker i Sverige: En idés väg från "andskap" till landskap. *Svensk Geografisk Årsbok 1991*, pp. 83–96.

Hägerstrand, T. 1993. Samhälle och natur. *NordREFO*, 1993(1): 14–59.

Hägerstrand, T. 2000. Landskapet som filter. In: Skärbäck, E. and Skage, O.R. (eds). *Planering för landskap*. Alnarp: SLU Alnarp, pp. 118–130.

Hägerstrand, T. 2004. The two vistas. *Geografiska Annaler: Series B*, 86(4): 315–323. Posthumously published.

Hägerstrand, T. 2009. *Tillvaroväven*. Stockholm: Forskningsrådet Formas. Posthumously published.

Hägerstrand, T. and Lenntorp, B. 1974. Samhällsorganisation i tidsgeografiskt perspektiv. *SOU 1974:2*, pp. 221–232, bilaga 2.

Hannah, M. G. 2013. Attention and the phenomenological politics of landscape, *Geografiska Annaler: Series B*, 95(3): 235–250.

Henderson, G. 2003. What (else) we talk about when we talk about landscape. For a return to the social imagination. In: Wilson, C., and Groth, P. (eds). *Everyday America: Cultural Landscape Studies after J.B. Jackson*. Berkeley, CA: University of California Press, pp. 178–198.

Lenntorp, B. 1976. *Paths in Space-Time Environments: A Time-Geographic Study of Movement Possibilities of Individuals*. Lund: Lunds Universitet.

Lenntorp, B. 1999. Time-geography: At the end of its beginning. *GeoJournal*, 48: 155–158.

Lenntorp, B. 2004. Path, prism, project and population: An introduction. *Geografiska Annaler: Series B*, 86(4): 223–226.

Lenntorp, B. 2011. En inblick i den hägerstrandska vävens begreppsflora. In: Palm, J. and Wihlborg, E. (eds). *Sammanvävt: Det goda livet i vardagsforskningen*. Linköping: Tema teknik och social förändring. Linköpings Universitet, pp.123–137.

Mårtensson, S. 1979. *On the Formation of Biographies in Space-Time Environments*. Lund: Lunds Universitet.

Mels, T. (1999). *Wild Landscapes: The Cultural Nature of Swedish National Parks*. Lund: Lund University.

Mitchell, D. 1996. *The Lie of the Land: Migrant Workers and the California Landscape*. Minneapolis: University of Minnesota Press.

Mitchell, D. 2008. New axioms for reading the landscape: Paying attention to political economy and social justice. In: Wescoat, J.L., and Johnston, D.M. (eds). *Political Economies of Landscape Change: Places of Integrative power*. Dordrecht: Springer.

Olsson, G. 1998. Horror vacui. In: Gren, M., and Hallin, P.O. (eds). *Svensk kulturgeografi: En exkursion inför 2000-talet*. Lund: Studentlitteratur, pp. 117–132.

Olwig, K.R. 1996. Recovering the substantive nature of landscape. *Annals of the Association of American Geographers*, 86(4): 630–653.

Olwig, K.R. 2002a. *Landscape, Nature, and the Body Politic: From Britain's Renaissance to America's New World*. Madison: University of Wisconsin Press.

Olwig, K.R. 2002b: The duplicity of space: Germanic "Raum" and Swedish "rum" in English language geographical discourse. *Geografiska Annaler: Series B*, 84(1): 1–17.

Olwig, K.R. 2005. The landscape of "customary" law versus that of "natural" law. *Landscape Research*, 30(3): 299–320.

Olwig, K.R. 2017. Geese, elves, and the duplicitous, "diabolical" landscaped space of reactionary modernism: The case of Holgersson, Hägerstrand, and Lorenz. *GeoHumanities*, 3(1): 41–64.

Olwig, K.R., and Mitchell, D. 2007. Justice, power and the political landscape: From American space to the European Landscape Convention. *Landscape Research*, 32: 525–531.

Pred, A. 1981. Social reproduction and the time-geography of everyday life. *Geografiska Annaler, Series B*, 63(1): 5–22.

Qviström, M. 2003. *Vägar till landskapet: om vägars tidrumsliga egenskaper som utgångspunkt för landskapsstudier*. Alnarp: Sveriges lantbruksuniv.

Qviström, M. 2015. Putting accessibility in place: A relational reading of accessibility in policies for transit-oriented development. *Geoforum*, 58: 166–173.

Rose, G. 1993. *Feminism and Geography: The Limits of Geographical Knowledge*. Cambridge: Polity Press.

Sanglert, C.J. 2013. *Att skapa plats och göra rum: landskapsperspektiv på det historiska värdets betydelse och funktion i svensk planering och miljövård*. Lund: Lunds universitet.

Sauer, C. 1925. *The Morphology of Landscape*. Berkeley: University of California Press.

Schwanen, T. 2007. Matter(s) of interest: Artefacts, spacing and timing. *Geografiska Annaler: Series B*, 89(1): 9–22.

Setten, G. 2002. *Bonden og landskapet: historier om natursyn, praksis og moral i det jærske landskapet*. Trondheim.

Sollbe, B. 1991. Professorn från Moheda. Interview with Torsten Hägerstrand. In: Carlestam, G. and Sollbe, B. (eds). *Om tidens vidd och tingens ordning: Texter av Torsten Hägerstrand*. Stockholm, Byggforskningsrådet.

Sörlin, S. 1991. *Naturkontraktet: om naturumgängets idéhistoria*. Stockholm: Carlsson.

Sui, D. 2012. Looking through Hägerstrand's dual vistas: Towards a unifying framework for time-geography. *Journal of Transport Geography*, 23: 5–16.

Tilley, C. 1994. *A Phenomenology of Landscape: Places, Paths and Monuments*. Oxford: Berg.

Widgren, M. 2004. Can landscapes be read? In: Palang, H., Sooväli, H., Antrop, M. and Setten, G. (eds). *European Rural Landscapes: Persistence and Change in a Globalising Environment*. Boston: Kluwer Academic, pp. 455–465.

Wihlborg, E. 2005. Flexible use of time to overcome constraints: A time-geographical discussion about power and flexibility. *Home-Oriented Informatics and Telematics, IFIP*, 178: 1–14.

Wihlborg, E. 2011. Makt att äga rum: En essä om tidsgeografisk epistemologi. In: Palm, R. and Wihlborg, E. (eds). *Sammanvävt: Det goda livet i vardagsforskningen. En vänbok till Kajsa Ellegård*. Tema teknik och social förändring. Linköping: Linköpings Universitet, pp. 108–122.

Williams, D.C. 1953. The myth of passage. *Journal of Philosophy*, 84(15): 457–472.

Wylie, J. 2007. *Landscape*. Abingdon: Routledge.

Wylie, J. 2013. Landscape and phenomenology. In: Howard, P., Thompson, I., and Waterton, E. (eds). *Routledge Companion to Landscape Studies*. London: Routledge, pp. 54–65.

Index

Locators in **bold** refer to tables and those in *italics* to figures.